Solving Differential Equations
with
MAPLE ®
Release 4

David Barrow
Art Belmonte
Al Boggess
Jack Bryant
Tom Kiffe
Jeff Morgan
Maury Rahe
Kirby Smith
Mike Stecher

Texas A & M University

Brooks/Cole Publishing Company

I(T)P® An International Thomson Publishing Company

Pacific Grove • Albany • Belmont • Bonn • Boston • Cincinnati • Detroit
Johannesburg • London • Madrid • Melbourne • Mexico City
New York • Paris • Singapore • Tokyo • Toronto • Washington

Project Development Editor: *Gary W. Ostedt*
Editorial Associate: *Carol Benedict*
Marketing Team: *Jill Downey and Laura Caldwell*
Signing Sales Representative: *Ragu Raghavan*
Production Editor: *Mary Vezilich*
Cover Design: *Christine Garrigan*
Printing and Binding: *Edwards Brothers*

Brooks/Cole — Thomson Learning
10 Davis Drive
Belmont CA 94002-3098
USA

For information about our products, contact us:
Thomson Learning Academic Resource Center
1-800-423-0563
http://www.brookscole.com

For permission to use material from this text, contact us by
Web: http://www.thomsonrights.com
Fax: 1-800-730-2215
Phone: 1-800-730-2214

Printed in the United States of America

5

ISBN 0-534-34555-7

Contents

Introduction

A computer algebra system such as Maple V Release 4 can be a valuable adjunct in a course in ordinary differential equations. Such a system provides immediate graphical feedback to the student about the behavior of solutions. If the course is to include an introduction to the numerical approximation of solutions, the principal means by which many nonlinear equations are addressed, some software system is almost a necessity. Moreover, Maple's symbolic manipulation prowess can be used to improve accuracy and speed in computation, allowing the students to spend less time on routine but time consuming differentiation, integration, and simplification, and allowing them to allocate more time to modeling and analyzing results of the computations that they perform.

One problem with implementing such an approach is that the richness and power inherent in Maple is coupled with an equally high complexity of syntax. Notwithstanding Maple's excellent Help facility, there is a often a significant gap between knowing what manipulation one wants to perform and knowing which sequence of Maple commands is both effective and efficient to do it. It is also difficult to know what problems one can reasonably assign to students who have such a tool. Finally, for an instructor new to the use of computer algebra systems in such a differential equations course, there are questions about what parts of the course are amenable to computer supplementation and how to go about it.

One goal of this manual is to provide examples of sequences of Maple commands that we have found useful, in the precise differential equations context where we have used them in our classes. A second goal is to provide a set of exercises and projects that make full use of the power of Maple, since the problem sets that one encounters in many textbooks are necessarily of extremely limited complexity to allow for hand solution. A third goal, one ever in our minds, has been, in the midst of Maple syntax, to maintain the focus of the student on the significant mathematical ideas inherent in this rich and fascinating subject.

Those goals having been stated, we should also add that, while a careful reading of this manual will reward individuals with a significant body of mathematical knowledge in addition to Maple syntax, there has been no intent to make this document totally self contained. We envision this manuscript as only one part of the course, to be accompanied by lecture/demonstrations on the part of knowledgeable faculty, with readings and some problems selected from one of the many fine textbooks in the area.

Toward the first goal, the basic Maple commands dealing with differential equations are contained in Chapters 1, 3, 5, and 6. Chapter 1 introduces Maple's `dsolve` command and its `numeric` option for solving a single ordinary differential equation. Chapter 3 involves higher order differential equations, while Chapter 5 discusses the commands that are relevant to solving differential equations using the Laplace transform. Chapter 6 introduces the commands and techniques required to solve systems of ordinary differential equations.

Exercises that illustrate the use of the commands are given at the end of each chapter. In addition, Chapters 2, 4 and 7 provide some applications. Chapter 8 delves deeper into the numeric methods first introduced in Chapter 1. Chapter 9 gives an introduction to partial differential equations. Chapter 10 contains projects (more extended applications) that utilize Maple to simplify the computations and to provide some graphical insight.

It is hoped that these exercises and projects will provide some glimpse of the power of Maple (and computer algebra systems in general) in simplifying the computations and graphics that are often required in the practical use of mathematics.

Chapters 7 and 8 and the last section in Chapter 1 are written with more pedagogical goals in mind. Chapter 8 and the last section in Chapter 1 use Maple to help illustrate several standard techniques for numerically approximating solutions to differential equations. Of course, Maple's `numeric` option to the `dsolve` command usually makes it unnecessary to encode a numerical algorithm. However, an explanation of the basic numerical routines is important in its own right and helps to explain some of the background behind the `numeric` option of `dsolve`. Chapter 7 uses Maple's extensive plotting capabilities to show the phase portraits of simple two dimensional systems.

This manual assumes that the reader has been exposed to most of the Maple commands used in calculus. A quick review of these commands is contained in Chapter 0. Students new to Maple can find a more comprehensive introduction in other references (e.g., *CalcLabs with Maple V*).

This manual was written using Maple V Release 4 and then exported as LaTeX files.

Chapter 0

Maple Review

This chapter presents a rapid review of some basic Maple commands. It is not intended to be a self-contained tutorial for those who have never seen Maple. There are many good Maple references for those new to Maple; for example, *CalcLabs with Maple V*, Brooks/Cole Publishing, ISBN 0-534-25590-6.

0.1 Assignment Statements and Arithmetic

Maple can perform all the usual arithmetic operations such as +, −, *(multiplication), / (division), and ^ (exponentiation). For example, the following statement computes $(3.2^4 + 4(9.8))/34$.

```
>   (3.2^4+4*(9.8))/34;
```

$$4.236988236$$

Note that all Maple statements end in a semicolon. Here and throughout this manual, Maple statements are entered after the > prompt. Maple's output is given in the center of the page.

To assign a label to a number, use :=. For example, the following two Maple statements assign the label a to 32/17 and the label b to 2.3.

```
>   a:= 32/17;b:=2.3;
```

$$a := \frac{32}{17}$$

$$b := 2.3$$

Arithmetic operations can be performed on a or b; for example, the following statements compute ab^2 and $\sqrt{a+b}$ with $a = 32/17$ and $b = 2.3$.

```
>   a*b^2;sqrt(a+b);
```

$$9.957647058$$

$$2.045080180$$

1

To compute the decimal equivalent of $a = 32/17$, use `evalf`.

```
>  evalf(a);
```

$$1.882352941$$

It is important that the reader be able to distinguish an assignment statement `z:=3;`, with symbol `:=`, from an equation, `z=3;`. The assignment statement makes a global change in values throughout the worksheet, as you have seen above. On the other hand, entering the equation `z=3;` only prints the equation on the terminal, but does not change any values at all.

```
>  z=3;z; # An equation.
```

$$z = 3$$

$$z$$

```
>  z:=3;z; # An assignment.
```

$$z := 3$$

$$3$$

The following statement unassigns the label a (removes the value 32/17 from a). (One can remove *all* assignments previously made in a worksheet with the command `restart;`.)

```
>  a:='a';
```

$$a := a$$

Labels can be assigned to any statement. The following statement assigns the label f to the expression $x^2 + 3x - 1$.

```
>  f:=x^2+3*x-1;
```

$$f := x^2 + 3x - 1$$

Note the use of `*` for multiplication—an error message will result if you type `3x;` without it. Algebraic operations can now be performed on f. For example, to expand f^2, i.e., $(x^2 + 3x - 1)^2$, type

```
>  expand(f^2);
```

$$x^4 + 6x^3 + 7x^2 - 6x + 1$$

0.2   Expressions versus Functions

In the preceding section, we assigned to f an *expression* involving x. To find the value of this expression for a specific value of x, such as $x = 2$, we cannot use the notation $f(2)$. Instead, in order to substitute a value, such as $x = 2$, into an expression $f := x^2 + 3x - 1$, the `subs` command must be used.

```
>  subs(x=2,f);
```

To use the more familiar function notation $f(2)$ for evaluating f at $x = 2$, f must first be defined as a *function*. To define f as a function, enter

```
>  f:=x->x^2+3*x-1;
```

$$f := x \to x^2 + 3x - 1$$

Note the arrow syntax, which is typed as a minus sign *immediately* followed by a greater than sign. (Intervening spaces will cause an error, since Maple interprets the isolated symbol as a >.) Now $f(2)$ can be evaluated using the notation $f(2)$,

```
>  f(2);
```

$$9$$

as can $f(x + h)$,

```
>  f(x+h);
```

$$(x + h)^2 + 3x + 3h - 1$$

Changing between the two forms is easy. An expression in the variable x can be converted to a function in the variable x by using the `unapply( ,x)` command. When a function f is evaluated, e.g., $f(x)$, the result is again an expression .

```
>  f:=x^3+2*sin(x);f:=unapply(f,x);f(x);
```

$$f := x^3 + 2\sin(x)$$

$$f := x \to x^3 + 2\sin(x)$$

$$x^3 + 2\sin(x)$$

Functions of more than one variable can be defined in an analogous manner. The following statement defines a Maple function (called *vol*) which represents the volume of a cylinder of radius r and height h.

```
>  vol:=(r,h)->Pi*h*r^2;
```

$$vol := (r, h) \to \pi h r^2$$

Note that π is typed as `Pi` . The volume of the cylinder of radius 3 and height 6 can be computed as follows.

```
>  vol(3,6);evalf(");
```

$$54\pi$$

$$169.6460033$$

The first statement evaluates *vol*(3, 6) and the second statement converts the number to a decimal. The quote or ditto " refers to the output of the *chronologically* immediately preceding Maple command. (Since it is possible to jump around in the windows worksheet environment, the user may choose not to execute the instructions sequentially. For this reason, the " may not refer to the command immediately above it in the worksheet. It is good Maple practice to place commands using the " as second commands on the same line, so that the antecedent is unambiguous.)

The reader should note that the subs command that is used with expressions is only a replacement. The following command looks for occurrences of x in the expression and replaces them by whatever object is on the right hand side of the equation that contains x. Simplifications are *not* automatic.

```
>   subs(x=ln(t),exp(x));simplify(");
```
$$e^{\ln(t)}$$

$$t$$

0.3   Plots

To plot $x^2 + 3x - 1$, first enter it as a Maple expression;

```
>   f:=x^2+3*x-1;
```
$$f := x^2 + 3\,x - 1$$

and then enter the following command, which plots f over the interval $-4 \le x \le 1$.

```
>   plot(f,x=-4..1);
```

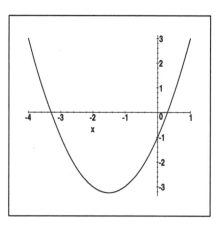

In Release 4, plots can either be inline (in the worksheet itself) or in a separate plot window. Clicking the mouse in the plot will change the menu items above the plot and bring up the options that apply to the plot. These plot menu items can be used to change the various aspects of a plot – e.g., style, axes, projection, etc.

As another example, we plot the expression

> `g:=1/x;plot(g,x=-2..2);`

$$g := \frac{1}{x}$$

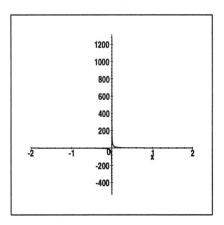

Note that Maple scales the y-axis to fit the values that come from the plotted x's. Since one of the points has a very large ordinate, the graph is scaled so that almost all the information is hidden. Whenever a graph has a vertical asymptote, we need to restrict the scale of the y-axis. To view the piece of the graph in the range $-6 \le y \le 6$, type the command

> `plot(g,x=-2..2,y=-6..6);`

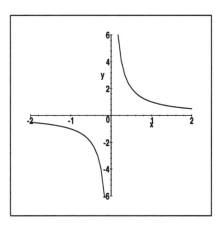

Both graphs can be viewed on one coordinate system by enclosing both f and g in braces { }.

> `plot({f,g},x=-4..2,y=-6..6);`

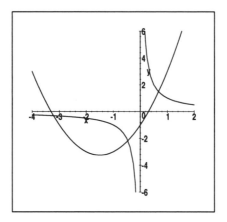

In Release 4, you can even choose the linestyle and color used to draw individual curves. The expressions must be input as a list, i.e., with square brackets [ ], since order is not preserved when objects are enclosed in set braces { }; and Maple must know to which expression the option applies.

```
>  myoptions:=linestyle=[0,2],color=[red,blue]:
```

```
>  plot([f,g],x=-4..2,y=-6..6,myoptions);
```

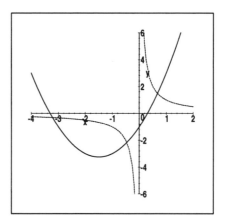

If f is entered as a Maple function rather than as a Maple expression (i.e., f:=x->x^2+3*x-1;), then the plot command should be typed either as plot(f(x),x=-4..1); or as plot(f,-4..1);. Mixtures of syntax, such as plot(f(x),-4..1); or plot(f,x=-4..1);, when f is a Maple function, will not work. Likewise, if f is a Maple expression, then the commands plot(f(x),x=-4..1); and plot(f,-4..1); will not work ($f(x)$ will not make sense unless f is defined as a function).

Two other useful graphics commands are `textplot` (used to annotate plots) and `display` (used to superimpose two or more separate plots). These are available in Maple's `plots` package, which is loaded via `with(plots):`. Here is an example.

```
>  with(plots):

>  p1:=plot(sin(2*x)+cos(3*x),x=-5..5):

>  p2:=textplot([3,-1.5,'Minimum']):

>  display([p1,p2]);
```

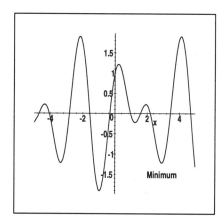

Maple can also work with expressions which are defined differently on different intervals. For example,

```
>  gee:=piecewise(x<0,0,x<1,x^2,x<2,1,x<3,0.5);
```

$$gee := \begin{cases} 0 & x < 0 \\ x^2 & x < 1 \\ 1 & x < 2 \\ .5 & x < 3 \end{cases}$$

The expression can be plotted as usual, where we use the option `discont=true` to avoid a nearly vertical line segment at the discontinuity.

```
>  plot(gee,x=-1..4,discont=true);
```

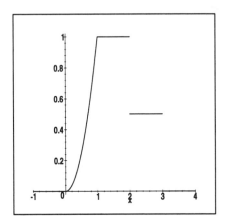

0.4   Solving Equations

Equations can be given labels in the same way that expressions can be given labels. For example, the command

> `eq:=x^3-2.0*x^2+x-3.0=0;`

$$eq := x^3 - 2.0\,x^2 + x - 3.0 = 0$$

assigns the label *eq* to the equation $x^3 - 2x^2 + x - 3 = 0$. To solve this equation, use the `solve` command.

> `solve(eq,x);`

$$-.08727970515 - 1.171312111\,I, \; -.08727970515 + 1.171312111\,I, \; 2.174559410$$

Note that there is one real root (≈ 2.175) and two complex roots (here, $I = \sqrt{-1}$). Also, the equation *eq* is entered with decimal coefficients (i.e. 2.0 is the coefficient of x^2) and so approximate decimal roots are returned. If the equation is entered with non-floating point coefficients (i.e., if `2*x^2` and `-3` are entered instead of `2.0*x^2` and `-3.0`), then `solve` will return exact roots (which in this case will be very messy).

We can observe the real root by plotting the expression $x^3 - 2x^2 + x - 3$ appearing on the left side of the equation *eq*. To do this efficiently in Maple, type the command

> `plot(lhs(eq),x=0..3);`

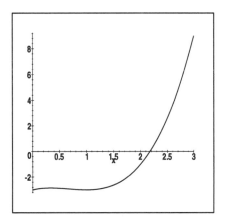

The command lhs gives the *left-hand side*. Observe that there is a root at about $x = 2.2$.

As mentioned earlier, solve tries to find an exact root of an equation (returning exact roots if the coefficients are specified using non-floating point values). However, many equations cannot be solved exactly. In this case an approximate solution can be found by using the fsolve command. Usually this command is used in conjunction with a plot. For example, to find the root of the equation $x^7 + 3x^4 + 2x - 1$, first enter this expression in Maple and then plot.

```
>   f:=x^7+3*x^4+2*x-1;plot(f,x=-1..2,y=-6..6);
```
$$f := x^7 + 3x^4 + 2x - 1$$

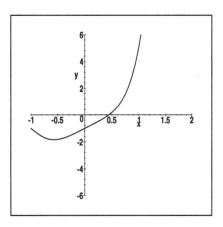

As the plot shows, f has a root between $x = 0$ and $x = 1$. This root can be found with the following command

```
>   fsolve(f=0,x,x=0..1);
```
$$.4414177090$$

Note that the equation comes first (`f=0`), the variable to solve for comes next (`x`), and the interval that contains the solution comes last (`x=0..1`).

Equations can be solved simultaneously by enclosing the equations in set braces. For example, to solve the equations

> `eq1:=2*r+3*s=5;eq2:=r-4*s=2;`

$$eq1 := 2\,r + 3\,s = 5$$

$$eq2 := r - 4\,s = 2$$

for r and s, enter the command

> `solve({eq1,eq2},{r,s});`

$$\{r = \frac{26}{11},\ s = \frac{1}{11}\}$$

Sometimes it is useful to assign a label to the solutions (such as *sol*) for future reference.

> `sol:=solve({eq1,eq2},{r,s});`

$$sol := \{r = \frac{26}{11},\ s = \frac{1}{11}\}$$

The variable *sol* contains two elements, corresponding to the two roots. Each element can be accessed individually in Maple by typing `sol[1]` or `sol[2]`. It is important to realize that r and s have not been assigned the values 26/11 and 1/11 (if you type r; or s; the values 26/11 or 1/11 will not appear as output). The `assign` statement can be used to assign these values to r and s.

> `assign(sol);`

Now the values of r and s have been assigned.

> `r;s;`

$$\frac{26}{11}$$

$$\frac{1}{11}$$

0.5 Differentiation

The differentiation command is either `diff` or D depending on whether a Maple expression or a function is being differentiated. To differentiate the *expression*

> `f:=x^2*sin(x);`

$$f := x^2 \sin(x)$$

type

> `diff(f,x);`

$$2\,x \sin(x) + x^2 \cos(x)$$

Note that the expression to be differentiated (in this case f) is given first in the command, and then the variable of differentiation (in this case x) comes next.

If f is entered as a function,

> `f:=x->x^2*sin(x);`

$$f := x \rightarrow x^2 \sin(x)$$

then `D(f)` returns the derivative.

> `D(f);`

$$x \rightarrow 2\,x \sin(x) + x^2 \cos(x)$$

`D(f)` is itself a function (note the arrow in the output of `D(f)` indicating that it is a function). So to evaluate the derivative of f at $x = 2$, type

> `D(f)(2);`

$$4 \sin(2) + 4 \cos(2)$$

On the other hand, `diff(f,x)` is an expression. To evaluate the derivative at $x = 2$ using this syntax requires the more cumbersome `subs` command.

Higher derivatives can also be computed. The second derivative of the function f can be computed by typing `(D@@2)(f);` the third derivative can be computed by typing `(D@@3)(f);` and so forth. (Here, we have used Maple's `@`, the composition operator.) Second and higher derivatives of expressions can be formed using either an explicit sequence of x's or the `$` sequence operator. (Note that while `x$2` works fine at the command prompt, inline mathematics in Release 4 will not accept the notation.)

> `gg:=x^2*sin(x);diff(gg,x,x);diff(gg,x$2);`

$$gg := x^2 \sin(x)$$

$$2 \sin(x) + 4\,x \cos(x) - x^2 \sin(x)$$

$$2 \sin(x) + 4\,x \cos(x) - x^2 \sin(x)$$

Partial derivatives can also be computed. For example, if $f = x^2 y^3$ is defined as an expression (say as `f:=x^2*y^3;`), then `diff(f,x)` and `diff(f,y)` are the partial derivatives of f with respect to x and y, respectively. On the other hand, if f is defined as a Maple function (via `f:=(x,y)->x^2*y^3;`), then `D[1](f)` and `D[2](f)` represent the partial derivative functions f_x and f_y, respectively. The variable with respect to which we are differentiating is specified by *position* (1 or 2 in the preceding case). In this regard, we recommend *alphabetizing* a function's input variables (e.g., (x, y)) so as to impose a natural order.

To get the implicit derivative, Release 4 has a new command:

> `gg:=x^2+y^3=1;Dy=implicitdiff(gg,y,x);`

$$gg := x^2 + y^3 = 1$$

$$Dy = -\frac{2}{3}\frac{x}{y^2}$$

0.6   Integration

The command for integrating an expression is `int`. For example, to integrate the expression

> ` f:=x^2*sin(x);`

$$f := x^2 \sin(x)$$

enter the command

> ` int(f,x);"+C;`

$$-x^2 \cos(x) + 2 \cos(x) + 2\,x \sin(x)$$

$$-x^2 \cos(x) + 2 \cos(x) + 2\,x \sin(x) + C$$

Note that Maple does not insert the constant of integration. You must enter it yourself if you want a constant of integration in the output.

It is recommended that you first display the integral using `Int` to check for errors in typing and then evaluate the integral using `value(")`; For example, the above integral can be evaluated using this method as follows. (`Int` is an *inert* command; i.e., its function is to display, rather than to evaluate.)

> ` Int(f,x);value(");`

$$\int x^2 \sin(x)\, dx$$

$$-x^2 \cos(x) + 2 \cos(x) + 2\,x \sin(x)$$

Definite integrals can be evaluated by inserting the limits into the `Int` command. For example, the integral $\int_0^1 x^2 \sin(x)\, dx$ is evaluated by typing

> ` Int(f,x=0..1);value(");`

$$\int_0^1 x^2 \sin(x)\, dx$$

$$\cos(1) + 2 \sin(1) - 2$$

The above integral is computed in Maple by the Fundamental Theorem of Calculus (i.e., an antiderivative is computed and then evaluated at the endpoints). If an antiderivative cannot be found, then a numerical approximation of the integral can be evaluated using `evalf(")`; instead of `value(")`; For example,

> ` Int(sqrt(x^5+1),x=0..1);evalf(");`

$$\int_0^1 \sqrt{x^5 + 1}\, dx$$

$$1.074669189$$

In the above example, no simple closed formula for the antiderivative exists (the `value(")`; command will not evaluate it).

The following example puts together several Maple commands to compute an area.

Example

Find the area bounded by the x-axis and the curve given by the following expression.

```
>   f:=-0.128*x^3+1.728*x^2-5.376*x+1.864;
```
$$f := -.128\,x^3 + 1.728\,x^2 - 5.376\,x + 1.864$$

First, f is plotted over an interval that shows all points where f crosses the x-axis (in this case the interval $-2 \le x \le 10$ will do).

```
>   plot(f,x=-2..10);
```

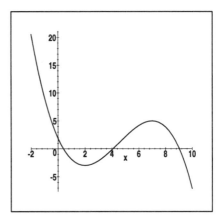

In this instance, since f is a polynomial, all the roots of f can be found with fsolve.

```
>   rt:=fsolve(f=0,x);
```
$$rt := .3955411814,\ 4.079363996,\ 9.025094823$$

From the graph, f is negative from $x = rt[1]$ to $x = rt[2]$ and f is positive from $x = rt[2]$ to $x = rt[3]$. Therefore, the following integral represents the desired area.

$$-\int_{rt[1]}^{rt[2]} f\,dx + \int_{rt[2]}^{rt[3]} f\,dx.$$

This can be computed in Maple as follows.

```
>   -Int(f,x=rt[1]..rt[2])+Int(f,x=rt[2]..rt[3]);value(");
```

$$-\int_{.3955411814}^{4.079363996} -.128\,x^3 + 1.728\,x^2 - 5.376\,x + 1.864\,dx$$
$$+ \int_{4.079363996}^{9.025094823} -.128\,x^3 + 1.728\,x^2 - 5.376\,x + 1.864\,dx$$

$$23.12794857$$

The area of interest is about 23.1 square units.

0.7   Taylor Approximations

The command to compute the Taylor polynomial of an expression is `taylor`. For example, to compute the Taylor polynomial of degree 2 of the expression

```
>   f:=sin(x)*exp(x);
```

$$f := \sin(x)\, e^x$$

about the point $x = -1$, enter the following command.

```
>   t2:=taylor(f,x=-1,3);
```

$$t2 :=$$
$$-\sin(1)\, e^{(-1)} + (-\sin(1)\, e^{(-1)} + \cos(1)\, e^{(-1)})\, (x + 1) + \cos(1)\, e^{(-1)}\, (x + 1)^2 + O((x + 1)^3)$$

Note that the exponential function is entered as `exp(x)`. In this case, we have labeled the Taylor expansion $t2$. The term $O((x+1)^3)$ indicates that the error is third order in $(x+1)$. Note that the integer at the end of the `taylor` command (in this case 3) is the same as the order of the remainder. Thus, this integer is one greater than the order of the desired Taylor polynomial. So, for example, the command to compute a fifth order Taylor expansion is `taylor(f,x=-1,6);`.

For many applications (such as a plot), the remainder term must be removed. To do this, use the `convert` command with the `polynom` option (for "polynomial") as follows.

```
>   t2:=convert(t2,polynom);
```

$$t2 := -\sin(1)\, e^{(-1)} + (-\sin(1)\, e^{(-1)} + \cos(1)\, e^{(-1)})\, (x + 1) + \cos(1)\, e^{(-1)}\, (x + 1)^2$$

Now $t2$ contains just the Taylor polynomial of degree 2 (without the remainder). A plot of both the Taylor polynomial and the original expression can be obtained with the following command.

```
>   plot({f,t2},x=-3..1);
```

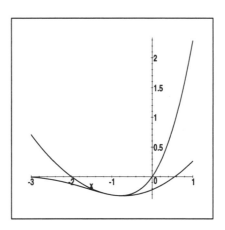

0.8 Exercises

1. The sequence operator $.

   (a) The second derivative of a Maple expression f is found by the command

   > `diff(f,x$2);`

   Find out what `x$2` is.

   (b) Find out what `$1..4` is. Next find out what `x."` is, where the `"` indicates the last result that Maple performed, in this case the `$1..4`. The dot between the x and the sequence `$1..4` is the concatenation operator.

2. Apply the command `convert( ,sincos)` to $1 + \tan^2 x$ and to $\tanh x$.

3. Plot $x^{2/3}$ over the interval $-2 \le x \le 2$. Try both

   > `x^(2/3);`

   and

   > `surd(x^2,3);`

   and explain why the first form doesn't work.

4. Numbers can be stored in Maple either as exact values or as floating point decimals.

   (a) Write the exact value `sqrt(2)`, and find out how many pieces it is stored in by using the command `nops(")`.

   (b) Find each piece of $\sqrt{2}$ by `op(1,sqrt(2))`, `op(2,sqrt(2))`.

   (c) Next, convert the exact value `sqrt(2)` into a floating point decimal using `evalf`.

   (d) Find out how many pieces a decimal number is stored in and what the pieces are.

5. Discover the properties of `expand`:

   (a) Expand $(x^2 - 1)^5$.

   (b) Expand $\sin(a + b)$.

   (c) Expand $\dfrac{a + b}{c}$.

   (d) Expand $\log(ab)$.

6. Write $x^3 + x^2 + x + 1$ in terms of powers of $x - 1$ using the `taylor` command.

7. Factor out $\sin x$ from $e^{2x} \sin x + \sin x \cos x + 4 \sin x$ by using `collect( ,sin(x))`.

8. Compute

   > `aaa:=taylor(x^2+x+4,x=2);`

   Now expand $aaa/(x - 2)$. What would you expect to get? Why don't you get it? Now try converting aaa to a polynomial, followed by a division by $(x - 2)$ and expand.

9. Send Maple back to school to learn `psqrt`, which is a `readlib` function. Now apply `sqrt` to x^2+2x+1 and `simplify`. Next apply `psqrt` to the same argument.

10. Determine the difference between `simplify` and `simplify( ,exp)`. Look in the index and find expressions that are treated differently by the two.

11. Tell Maple to assume that n and m are integers, and show that $\int_0^{2\pi} \sin nx \cos mx \, dx = 0$

12. Examine the Release 4 interface, and insert the equation $E = mc^2$ as inline mathematics in a text area by using the Σ button.

13. (Boggess) Insert an inline matrix into text by creating a matrix `matrix(2,2,[a,b,c,d])` at a Maple prompt, then mark and cut the matrix, use the Σ button to enter online math, and paste in the matrix.

14. (Belmonte) Use a null string ` ` to form the Maple expression `a<` `. Now form the inline mathematics expression $a < x < b$ by placing two inline math boxes side by side.

15. Solve the equation $x^4-4 = 0$ in three different ways: use `solve`, `fsolve`, and `fsolve( ,x,complex)` and compare your answers.

16. Use a hand calculator to compute $\ln(-2)$ and ask Maple to find the floating point decimal value of the same argument.

17. Plot the parametric curve $x = \sin 2t$, $y = \cos 2t$ for $0 \le t \le \pi$.

Chapter 1

Differential Equations and Maple

This chapter covers the use of the `dsolve` and `DEtools[DEplot]` commands. The `dsolve` command is used to generate an explicit solution to a differential equation with or without initial conditions. If this is not possible or desirable, for whatever reasons, this command has a `numeric` option which can be utilized if a numeric initial condition is specified. This option causes `dsolve` to construct a Maple procedure which can be used to approximate the actual solution. The second command, `DEplot`, is used to display the direction field of a differential equation, either with or without a solution, to an initial value problem. The second of these commands is accessed by first loading the `DEtools` library package. We also demonstrate how to plot the numerical approximations to solutions that are generated by the `numeric` option of the `dsolve` command.

The syntax of these two commands is discussed below. After that discussion several examples are worked out.

The chapter concludes with some exercises, whose difficulty varies from routine to challenging.

1.1   Maple Commands

- `with(DEtools)`: Use this command to load the `DEtools` library. If this library is not loaded, Maple will not recognize the command `DEplot`. (You will be able to see when Maple does not recognize the command because it will parrot back the command, i.e., it will print out the word `DEplot`.) It is, however, possible to use this command directly by use of the command `DEtools[DEplot]`.

- `dsolve`
  CALLING SEQUENCE:
  `dsolve(deqns,vars)`
  `dsolve(deqns,vars,eqns)`
  PARAMETERS:
  deqns—ordinary differential equation in vars, or set of equations and/or initial conditions
  vars—variable or set of variables to be solved for
  eqns—optional equations of the form keyword=value

- `DEplot`
  CALLING SEQUENCES:
  `DEplot(deqns,vars,trange,inits,eqns,...)`

```
DEplot(deqns,vars,trange,inits,xrange,yrange,eqns)
```
PARAMETERS:
deqns—differential equation(s)
vars—the names of the variables
trange—range of the independent variable
inits—initial conditions
xrange—range of a dependent variable
yrange—range of a second dependent variable
eqns—use Maple's `Help` facility to learn about available options

1.2 Examples

1.2.1 dsolve

The differential equation we are going to solve is:

$$\frac{dy}{dx} + 3y = x + e^{-2x} .$$

```
>   diffeq:=diff(y(x),x)+3*y(x)=x+exp(-2*x);
```

$$diffeq := (\frac{\partial}{\partial x} y(x)) + 3\,y(x) = x + e^{(-2\,x)}$$

Note: the unknown solution $y(x)$ must explicitly show its dependence on the independent variable, in this case x. *This dependence must be shown not only in the derivative, but everywhere that y occurs.*

```
>   sol:=dsolve(diffeq,y(x));
```

$$sol := y(x) = \frac{1}{3}\,x - \frac{1}{9} + e^{(-2\,x)} + e^{(-3\,x)}\,\_C1$$

Notice the *_C1* term. This is Maple's notation for an arbitrary constant. The *underscore* (_) is part of the symbol for the constant; it does not indicate subtraction. The solution is normally written as

$$y(x) = \frac{1}{3}x - \frac{1}{9} + e^{-2x} + Ce^{-3x} .$$

If we wish to find the solution to the differential equation which satisfies a particular initial condition, say $y(1) = 2$, then `dsolve` is used as follows. (Here we have suppressed the output of the first command by ending the Maple command with a colon. When the output does not give additional information, omitting output can make the worksheet clearer.)

```
>   inits:=y(1)=2:
```

```
>   sol:=dsolve({diffeq,inits},y(x)):sol:=simplify(");
```

$$sol := y(x) = \frac{1}{3}\,x - \frac{1}{9} + e^{(-2\,x)} + \frac{16}{9}\,e^{(-3\,x+3)} - e^{(-3\,x+1)}$$

Note: the differential equation and the initial condition **must** *be written as a set (i.e., enclosed by braces) in the first argument of the* `dsolve` *command.* If this is not done, Maple treats the initial condition as an unknown to be solved for.

Let's verify that the function $y(x)$ really is a solution by substituting it into the differential equation. The `subs` command is only a replacement. Since it conveys so little information, it is usually followed by a colon to suppress output. The derivatives are not actually taken until we execute the `simplify;` command.

```
>   subs(sol,diffeq):simplify(");
```
$$x + e^{(-2x)} = x + e^{(-2x)}$$

Since the equation is clearly valid, we know that our solution really does satisfy the differential equation. The following command plots the solution.

```
>   plot(rhs(sol),x=0..3);
```

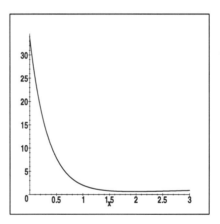

1.2.2 DEplot

Maple's ability to find an analytic expression for the solution to a differential equation is impressive, but the reader needs to know that there are many differential equations for which there are no known analytic representations of the solutions. Other commands are sometimes necessary to compute and interpret the solutions. One of Maple's tools is a command which can be used to construct the direction field of a differential equation, and, if desired, simultaneously sketch a numerical approximation of the solution to a particular initial value problem. The next example demonstrates the command `DEplot`.

The direction field for the differential equation (**Caveat: in standard form!**)

$$\frac{dy}{dx} = x + \sin y$$

is plotted as follows:

```
>   with(DEtools);diffeq:=diff(y(x),x)=x+sin(y(x));
```

[*DEnormal, DEplot, DEplot3d, Dchangevar, PDEchangecoords, PDEplot, autonomous, convertAlg, convertsys, dfieldplot, indicialeq, phaseportrait, reduceOrder, regularsp, translate, untranslate, varparam*]

$$diffeq := \frac{\partial}{\partial x} y(x) = x + \sin(y(x))$$

```
>  DEplot(diffeq,y(x),x=-3..3,y=-2..2);
```

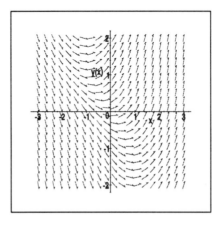

```
>   inits:=[0,-1],[0,0],[0,1]:
>   DEplot(diffeq,y(x),x=-3..3,{inits},y=-2..2);
```

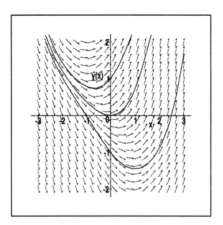

Notice that the range of the dependent variable is specified after the initial conditions. If we wish to plot just the solutions to the given initial conditions, the option `arrows=NONE` is used.

```
>   DEplot(diffeq,y(x),x=-3..3,{inits},y=-2..2,arrows=NONE);
```

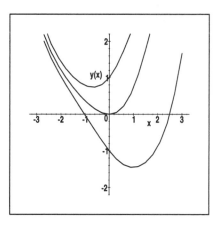

1.2.3 dsolve/numeric

Let's try to find an exact solution to this problem.

```
>   inits:=y(0)=1:
```

```
>   dsolve({diffeq,inits},y(x));
```

When this command is used, Maple's output is nonexistent, indicating that it cannot construct an analytic solution to this problem. In this case, an approximate solution can be found by using the numeric option, as in the following example.

```
>   sol:=dsolve({diffeq,inits},y(x),numeric);
```

$$sol := \mathbf{proc}(rkf45\_x) \ldots \mathbf{end}$$

To compute the solution at $x = 1$, enter sol(1);.

```
>   sol(1);
```

$$[x = 1, \, y(x) = 2.412875596843171]$$

When the numeric option of dsolve is activated, Maple constructs a procedure which implements an algorithm that approximates a numerical value of the solution at a prescribed value of the independent variable. The second command, sol(1);, caused Maple to use the previous procedure to compute an approximate value of the solution to the initial value problem at $x = 1$.

To see a plot of this numerical procedure, the Maple command odeplot *is used. To access this command, the Maple package* plots *is loaded.*

```
>   with(plots):odeplot(sol,[x,y(x)],-1..1);
```

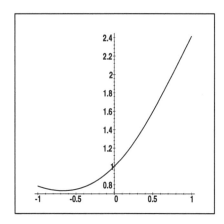

Notice the syntax of the `odeplot` command. It is *crucial* that the variables in the list [x,y(x)] be the same as those used when the Maple procedure *sol* was constructed. In particular, if the `odeplot` command is changed so that x is replaced by t, the coordinate axes will appear; but no curve will be generated! (Not all problems have x as the independent variable.)

There are various options available with this command. Peruse the `Help` facility to learn about them. Click on "Help," then "Topic Search," then "Apply." Alternatively, type `?plot,options` at a Maple prompt.

1.3   Euler's Method Approximation

In the previous two subsections, you have seen applications of `DEplot` and `dsolve/numeric`. These commands are related in an interesting way. In the `DEplot` direction field graphs, you saw arrows, which are tangent to the trajectories followed by solutions to the first order differential equation. You also observed that by following the arrows, one can trace out approximate trajectories. The `dsolve/numeric` command also produces trajectories. What is the actual procedure that is followed to trace out these paths?

Both `DEplot` and `dsolve/numeric` use variants of a moderately complicated algorithm due to Runge and Kutta. We will discuss their method at length in the chapter on numerical methods. However, in this section we introduce a much simpler approximation method due to Euler. Euler's method amounts to simply following the arrows in the direction field plot. This method, while not very accurate, is easy to understand and is implementable either with a hand calculator (for a few steps) or by Maple. The method will be executed first as a simple loop, and later as a Maple procedure that we write. The Euler method is also available as an option in `dsolve/numeric`. The interested reader can look in Help for the Topic `dsolve,classical` or glance ahead to the chapter on Numeric Methods.

Throughout this section we will construct numerical approximations to the initial value problem:

$$\frac{dy}{dx} = x + \sin y, \quad y(0) = 1. \tag{1.1}$$

```
>   f:=(x,y)->x+sin(y);inits:=y(0)=1;
```
$$f := (x, y) \to x + \sin(y)$$

$$inits := y(0) = 1$$

Notice that the function above is an arrow-defined function and not a Maple expression. When the various numerical algorithms are constructed, terms of the form $f(x_i, y_i)$ will be needed. They are much easier to enter via this notation than by use of the Maple subs command.

The first Maple procedure uses an Euler method to construct approximations to the solution of the differential equation:

$$\frac{dy}{dx} = f(x, y), \ y(firstx) = firsty.$$

The derivative of $y(x)$ is, at least for small h, approximately equal to $(y(x + h) - y(x))/h$. Therefore, the above differential equation can be approximated by the following equation

$$\frac{y(x + h) - y(x)}{h} \approx f(x, y(x))$$

which can be rewritten as

$$y(x + h) \approx y(x) + hf(x, y(x)).$$

The idea behind Euler's method is first to let $x = x_0$ and $y = y(x_0)$ be the x and y values given in the initial condition ($x_0 = 0$ and $y(x_0) = 1$ in the above example). Then the above equation computes the approximate y-value at $x_1 = x_0 + h$, i.e., $y(x_1) = y(x_0) + hf(x_0, y(x_0))$. Replacing x_0 by x_1 in the above equation yields an approximate y-value at $x_2 = x_1 + h$, i.e., $y(x_2) = y(x_1) + hf(x_1, y(x_1))$, and so forth.

To iterate this algorithm, we use the **for k from** *start* **to** *finish* **do .. od** construction. The following sequence of Maple commands performs ten iterations of this procedure with step size $h = 0.1$ and therefore computes approximate values for the solution at the points $x = 0, 0.1, 0.2, \ldots, 0.9, 1.0$. These values are contained in a sequence of points named *eulerseq*.

```
> x:=0:y:=1:h:=0.1:  # initialize x and y and the step size

> eulerseq:=[x,y]; # input the initial conditions into eulerseq
```

$$eulerseq := [0, 1]$$

```
> for i from 1 to 10 do
  y:=evalf(y+h*f(x,y)):  # compute the new value of y
  x:=x+h:  # update the new value of x
  eulerseq:=eulerseq,[x,y]:  # add the new point
  od:

> x:='x':y:='y':h:='h':
```

To display the contents of the solution values in *eulerseq*, type the variable name.

```
> eulerseq;
```

[0, 1], [.1, 1.084147098], [.2, 1.182537584], [.3, 1.295094550], [.4, 1.421317990],
  [.5, 1.560202880], [.6, 1.710197269], [.7, 1.869227210], [.8, 2.034807112],
  [.9, 2.204233584], [1.0, 2.374833354]

To plot the approximation between $x = 0$ and $x = 1$, we could use plot([eulerseq]);.

The above sequence of Maple commands can be improved and put into the form of a procedure (which is like a subprogram or subroutine in other computer languages). The following procedure (called *MyEuler*) takes as inputs the function f that describes the differential equation; the initial x-value (called *firstx*); the final value of x (called *lastx*); the initial y-value (called *firsty*) and the step size. The output of the procedure will be the approximate solution value (the y-value) corresponding to $x = lastx$. After the procedure is finished running, we would still like to access the variable *eulerlist* which contains all the approximate solution values over the interval *firstx* $\leq x \leq$ *lastx*; and so *eulerlist* has been declared as global. The other variables used by the procedure, such as x, y, h, and *point*, are declared as **local** so that the procedure's use of these variables will not interfere with use of the same variables outside the procedure. Their values will not be available to Maple after the procedure is complete.

(The student should note that the proper way to enter a procedure is with Shift-Return, rather than just Return. Shift-Return will do a carriage return and back the cursor to the start of the next line; but all the commands remain in the same group and are executed at the same time, with just one keystroke.)

```
>   MyEuler:=proc(f,firstx,lastx,firsty,stepsize)
    local x,y,h,point,eulerseq,tempy;
    global eulerlist;
    y:=firsty;h:=stepsize;
    eulerseq:=NULL;
    for x from firstx by h to lastx do
    point:=[x,y];
    eulerseq:=eulerseq,point:tempy:=y;
    y:= evalhf(y+h*f(x,y));
    od;
    eulerlist:=[eulerseq];
    tempy;
    end:
```

The output of this procedure is a numerical approximation to the solution at the point *lastx*. Notice that we have used floating point decimals in our computation. We do not want Maple doing exact arithmetic ("ordinary fractions"). If the function f is at all complicated, exact arithmetic really slows Maple down. This is perhaps not apparent in this simple algorithm, but it is very obvious in more complicated situations. In fact, where speed is important, we can use the command evalhf, which causes Maple to do the floating point arithmetic using machine hardware, rather than software.

Let's use this procedure to estimate the value of the solution at $x = 10$. We use a step size of 0.1.

```
>   MyEuler(f,0,10,1,0.1);
```

$$50.08529485935338$$

```
>   with(plots):
```

```
>   euler1:=plot(eulerlist,0..10,0..55):
```

```
>   display([euler1]);
```

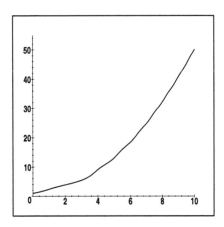

1.4 Exercises

1. For each of the following, use DEplot to draw the direction field associated with the differential equation. Based on the appearance of the direction field, how does the asymptotic behavior of solutions to these differential equations depend upon their initial values?

   (a) $y' = y(4 - y)$.

   (b) $y' = xe^{-2x} - 2y$.

   (c) $y' = x^2 + y^2$.

2. Compare the direction fields of the differential equations: $\dfrac{dy}{dx} = 1$, $\dfrac{dy}{dx} = x$ and $\dfrac{dy}{dx} = y$. What do you think the direction fields indicate about the behavior of solutions to these equations as x becomes very large?

3. For each of the following, use DEplot to show the direction field and the solution curve for the given initial condition.

   (a) $(\ln x)y' + y = \cot x$, $y(2) = 3$.

   (b) $(4 - x^2)y' + 2xy = 3x^2$, $y(-3) = 1$.

   (c) $y' - \dfrac{1}{x}y = \sqrt{x}$, $y(1) = 0$.

4. Recall that a differential equation is said to be in *standard form* if the coefficient of the highest order derivative is 1. Denote by $p(x)$ the coefficient of y in a first order linear differential equation in standard form. The quantity $e^{\int p(x)\, dx}$ is said to be an *integrating factor* for the equation. What are the integrating factors of the following differential equations?

   (a) $y' + x^2 y = \sin x$.

   (b) $y' + (\sin x)y = x$.

(c) $2xy' + e^x y = -x$.

(d) $x^2 y' + (\ln x)y = x$.

5. Use Maple to compute an integrating factor μ, written as an expression in the variable x; and solve the following first order linear equations. You can also use the following formula, if the equation is in standard form.

```
>  sol:=y(x)=(1/mu)*(y0*subs(x=a,mu)+int(subs(x=s,mu*f),s=a..x));
```

where $y0$ is the initial value at $x = a$, if there is one, or c if there is not; and f is the right hand side of the equation, written as an expression in x. You will see this formula again when we solve *systems* of simultaneous differential equations. Compare your work with the output from dsolve.

(a) $y' + y = xe^{-x} + 1$.

(b) $y' + \left(\dfrac{1}{x}\right) y = 2x, \ x > 0$.

(c) $xy' + (x+1)y = x, \quad y(\ln 2) = 1$.

(d) $xy' + 2y = 4x^2, \quad y(1) = 2$.

6. Find solutions to the following initial value problems. Before finding the solution with Maple, decide if you know a technique which you could use to solve the equation with pencil and paper. Then construct the solution with pencil and paper and compare it to the solution found with Maple. If necessary, use the numeric option of dsolve.

(a) $x' = 2 + \cos t, \quad x(\pi) = 1$.

(b) $x' = 2 + \cos x, \quad x(\pi) = 1$.

(c) $t^2 x' + x \sin t = 2 - t$.

(d) $yy' = e^x$.

7. Solve the differential equation $\dfrac{dy}{dx} = \begin{cases} -2y, & \text{if } -5 \le x \le 0 \\ 2y, & \text{if } 0 < x \le 5 \end{cases}$. (Note that in Release 4, dsolve will accept a piecewise expression. You still need to specify y as $y(x)$.)

8. Display the direction field of the equation $x^2 y' + 2xy = 1$ on $0 < x < \infty$, and predict the limiting behavior of solutions as $x \to \infty$.

9. For the differential equation $x^2 y' + 2xy = 1$, find all solutions which satisfy $y(2) = 2y(1)$. This type of problem is known as a boundary value problem.

10. Plot the solution of the differential equation for each of the given initial conditions. Put all of your plots for each set of initial conditions on the same graph.

(a) $x' = \dfrac{x}{10}, \quad x(0) = 1, \ 1.5, \ 2$.

(b) $x' = x - x^2, \quad x(0) = 0.5, \ 0.75, \ 1, \ 1.5, \ 2$.

(c) $x' = \sin x, \quad x(0) = 2.5, \ 3, \ \pi, \ 3.5$. This one is troublesome. Why?

11. Use `dsolve` on $y' = \pi y$ with π indicated in Maple as `pi`. Also try with π indicated as `evalf(Pi)`, and `convert(evalf(Pi),rational)`. Finally, try $y' = 0.1\pi y$. (Hint: use the `convert` command.) In Release 4, mixtures of floating point decimals and exact values such as π are not accepted in `dsolve`.

12. Use `dsolve` to show that if $y = \phi(x)$ is a solution of $y' + p(x)y = 0$, then $y = c\phi(x)$ is also a solution for any value of the constant c. Now use the "Help" facility to look up `DESol`, and reprove the result with `DESol`.

13. The equation $y' + p(x)y = q(x)y^n$ is nonlinear for values of n other than 0 or 1. However, it can be solved by converting it to a linear equation with the substitution $v = y^{1-n}$. Solve that new linear equation to get v, and then use v to recapture y. Use Maple to solve the equation $x^2 y' + 2xy - y^3 = 0$, $x > 0$. Translate the equation into Maple notation and assign it the label *old_ode*, then set `y:= v^(-1/2);`. Assign `new_ode:=old_ode;` and solve the resulting equation.

14. Consider the initial value problem

$$y' = \frac{y^2 + 2ty}{3 + t^2}, \quad y(1) = 2.$$

   (a) Use the `dsolve( , ,numeric)` command to find the value $y(2)$.

   (b) This equation is of *Bernoulli* type. (See your text.) Use this information to solve the equation and find the exact value $y(2)$.

   (c) Use `dsolve` to get an explicit formula for $y(2)$ with less work.

15. Solve the differential equation $y' + p(x)y = 0$, where $y(0) = 1$, $p(x) = \begin{cases} 2, & 0 \le x \le 1 \\ 1, & x > 1 \end{cases}$.

16. The $f(x, y)$ term in the differential equation $\dfrac{dy}{dx} = \dfrac{x^2}{y(1 + x^3)}$ is continuous everywhere except at $y = 0$ and $x = -1$. Find the domain of existence for the solution of the differential equation with the initial condition $y(0) = 1$.

17. In this chapter we have used Maple to solve differential equations of the form $\dfrac{dy}{dx} = f(x, y)$, and each time the $f(x, y)$ term was a Maple expression. What happens if this term is an arrow-defined function?

18. In the problems given above, we used mostly the built-in Maple commands to solve the differential equations. In your textbook, you will see examples of differential equations in which the variables separate. Solutions to such equations are often left in implicit form, rather than solving explicitly for y. For example, consider the equation $\dfrac{dy}{dx} = \dfrac{\sec^2 y}{1 + x^2}$. Solve this equation with the following Maple commands:

   ```
   >  eqn:=int(1/sec(y)^2,y)=int(1/(1+x^2),x)+C;
   ```

   Then check your solution using the Release 4 command `implicitdiff`:

19. Use Euler's method with step size $h = 0.1$ to approximate the solution to the initial value problem $y' = x - y^2$, $y(1) = 0$ at the points $x = 1.1, 1.2, 1.3, 1.4$, and 1.5.

20. Use Euler's method with step size $h = 0.05$ to approximate the solution to the initial value problem $y' = x - y^2, y(1) = 0$ at the points $x = 1.1, 1.2, 1.3, 1.4,$ and 1.5. (You'll have to compute intermediate points, but we throw them away.)

21. Use `dsolve` to get an exact solution to the initial value problem $y' = x - y^2$, $y(1) = 0$ in terms of Bessel functions. Do you know what Bessel functions are? Do you care? Substitute $x = 1.1, 1.2, 1.3,$ $1.4,$ and 1.5 into the right hand side of the solution, and use `evalf` to find floating point decimal values for the solution. Now compute the error between the exact answer and the two Euler approximants in the two previous problems. What happens to the error as we go farther away from the initial value of x? The step size in the second of the two previous problems is half as large as the step size in the first. What is the ratio for the corresponding errors at each x value?

22. Use `interface(verboseproc=2); print('dsolve/numeric');` to see the `dsolve` procedure.

23. **A Hysteresis Effect.** The procedure created by the numeric option in the `dsolve` command has an interesting feature. It remembers the last position for which it has calculated the value of a solution. If you then request the value of the solution at some other point, the procedure treats the last point as the initial point for the second request. This has some undesirable consequences. For example, use the `numeric` option of `dsolve` to solve the initial value problem

$$\frac{dx}{dt} = t^2 x, \; x(0) = 1.$$

Then, in this order, compute the values of $x(0), \; x(2), \; x(0), \; x(2), \; x(0).$

With Release 4, there is an option `startinit=true`, which will force the procedure always to start at the initial value. Repeat the process with this option.

24. An *isocline* for the differential equation $dy/dx = 2x^2 - y$ is a set of points in the xy-plane where all solutions have the same slope dy/dx. Use the `contourplot` command to sketch a few isoclines.

25. Recall that the exponential function is given in Maple by `exp`. Use `implicitdiff(eqn,y,x)` to find the implicit derivative for the equation $e^{xy} + y = x - 1$. Verify that the equation is an implicit solution to the differential equation by subtracting the implicit derivative from the right hand side of the differential equation and simplifying.

26. Use the `implicitplot` command on the rectangle $-4 \le x \le 4$ and $-4 \le y \le 4$ to sketch the solution curve $e^{xy} + y = x - 1$ for the above differential equation.

27. Explain what happens when you enter (as written) `(x+2)(x+4);` into a Maple command line. Why does it give this particular strange result, and what can be done to correct the error?

28. Compare the arrow plots given by `DEplot` for the equations $y' = (1 - x^2)/y^2$ and $y^2 y' = 1 - x^2$. Which one is correct? (This is a Maple "bug.") What form must an equation be in for `DEplot` to work? Does the form make a difference for `dsolve/numeric`? Try it for the initial value $y(1) = 1$.

Chapter 2

Applications

2.1 Art History

Someone brings you a painting that is supposedly 300 years old and painted by the 17th century Dutch master, Jan Vermeer. Some art experts claim that it is a recent forgery by the 20th century painter, Van Meegeren. What can be done to determine whether the painting is an original Vermeer? In the following case, art historians performed an analysis of the radioactive elements present in the painting. The pertinent facts are:

- White lead, a common pigment used for many hundreds of years in oil paintings, contains among other isotopes, Pb^{210}, a radioactive form of lead which has a half life of 22 years.

- An isotope of radium, Ra^{226}, has a half life of 1600 years and decays into Pb^{210}.

- Traces of radium are found in naturally occurring lead ore. The quantities of both radium and lead decrease exponentially over time, but eventually an equilibrium is established; and the ratio of disintegrations per minute of lead to disintegrations per minute of radium becomes one.

- The process of refining lead from ore removes almost all of the radium, but very little Pb^{210}, dramatically increasing the ratio, which then slowly drops to its original value of one.

- The value of the ratio at a given time can be used to establish bounds on the number of years since refining took place.

We will use these facts to determine whether or not the painting is a forgery. The idea is simple. By solving the differential equation for radioactive decay, and assuming that the pigment was refined 300 years ago, we can predict what the ratio of disintegrations per minute would have been at the time the refining took place. The ratio immediately after refining varies from mine to mine, but it should be somewhere between 0.18 and 140. If the predicted initial ratio is much larger than this range, one would strongly suspect that the ore was refined more recently and that the painting is indeed a forgery.

Let $w(t)$ be the number of atoms of Pb^{210} per gram of ordinary lead at time t, and $r(t)$ denote the number of disintegrations of Ra^{226} per minute per gram of ordinary lead at time t. If k is the decay constant for Pb^{210}, then

$$\frac{dw}{dt} = -kw + r(t), \ w(t_0) = w_0.$$

29

Since the painting is supposedly 300 years old, and the half life of Ra^{226} is 1600 years, the amount of Ra^{226} is essentially constant over the time interval with which we are concerned. Hence, $r(t)$ is essentially constant, too; and we denote this constant by r.

The solution to the above differential equation, with constant r, is:

```
>  restart:
```

```
>  eq1:=diff(w(t),t)=-k*w(t)+r;
```

$$eq1 := \frac{\partial}{\partial t}\, w(t) = -k\, w(t) + r$$

The expand command takes a fraction and puts each term of the numerator individually over the denominator.

```
>  dsolve({eq1,w(t0)=w[0]},w(t));expand(");
```

$$w(t) = \frac{r - \dfrac{e^{(-k\,t)}\left(e^{(-t0\,k)}\,r\,e^{(t0\,k)} - w_0\,k\right)}{e^{(-t0\,k)}}}{k}$$

$$w(t) = \frac{r}{k} - \frac{e^{(t0\,k)}\,r}{k\,e^{(k\,t)}} + \frac{e^{(t0\,k)}\,w_0}{e^{(k\,t)}}$$

We now simplify, using only the exponential simplification rules.

```
>  eq2:=simplify(",exp);
```

$$eq2 := w(t) = \frac{r}{k} - \frac{r\,e^{(k\,(-t+t0))}}{k} + w_0\,e^{(k\,(-t+t0))}$$

We now solve this equation for kw_0.

```
>  k*solve(eq2,w[0]);
```

$$\frac{k\,w(t) - r + r\,e^{(k\,(-t+t0))}}{e^{(k\,(-t+t0))}}$$

```
>  expand(");
```

$$\frac{e^{(k\,t)}\,k\,w(t)}{e^{(t0\,k)}} - \frac{e^{(k\,t)}\,r}{e^{(t0\,k)}} + r$$

```
>  est_initial_rate:=simplify(",exp);
```

$$est\_initial\_rate := k\,w(t)\,e^{(-k\,(-t+t0))} - r\,e^{(-k\,(-t+t0))} + r$$

Note that w is the number of atoms of radioactive lead per gram of ordinary lead. However, since k is the fraction of radioactive lead atoms that decay in one minute, it follows that kw is number of disintegrations of radioactive lead per minute, per gram of ordinary lead. Since the radium and radioactive lead are in equilibrium, this quantity must also be the number of disintegrations of radium per gram of ordinary lead, per minute, (loss of radioactive lead is counterbalanced by formation of new radioactive lead). It is labeled $est\_initial\_rate$ in the equation above.

The method applied to establish that the painting was a forgery was to use the above equation to predict the value of $kw(t_0)$ for the painting in question. The following table lists some of these values for Ra^{226}.

Material	Source	Disintegrations per minute per gram of lead of Ra^{226}
Ore concentrate	Oklahoma–Kansas	4.5
Crushed raw ore	S.E. Missouri	2.4
Ore Concentrate	S.E. Missouri	0.7
Ore Concentrate	Idaho	2.2
Ore Concentrate	Idaho	0.18
Ore Concentrate	Washington	140

From these values it seems that the value of kw_0 should vary between 0.18 and 140. If there is a huge difference between what our equation predicts and these values, then we have solid reason to believe the painting is not 300 years old.

The following table lists some of the supposedly forged paintings of Vermeer and approximate disintegration rates for Pb^{210} and Ra^{226} from these paintings.

Painting	Disintegrations per minute per gram of lead of Pb^{210}	Disintegrations per minute per gram of lead of Ra^{226}
Disciples at Emmaus	8.5	.8
Washing of Feet	12.6	.26
Woman Reading Music	10.3	.3

From the *est_initial_rate* equation, we see that we need the values of kw, r, k, and $t - t_0$. Remember that the relationship between the decay constant and the half-life of a material is

$$\text{half-life} = \frac{\ln 2}{k}. \tag{2.1}$$

Let's see what happens when we check the painting *Disciples at Emmaus*.

```
>   subs(t-t0=300,r=.8,w(t)=8.5/k,k=ln(2)/22,est_initial_rate):
```

```
>   evalf(");
```

$$98050.26122$$

It seems clear that 98,000 disintegrations per minute is considerably larger than we should expect. This leads to the conclusion that the painting *Disciples at Emmaus* is much younger than 300 years and is indeed a forgery.

References which will supply more details are:

- Braun, M., *Differential Equations and Their Applications*. Springer-Verlag, 1975

- Keisch, B., Feller, R. L., Levine, A. S., Edwards, P. R., "Dating and Authenticating Works of Art by Measurement of Natural Alpha Emitters," *Science*, 155(March, 1967), 1238–1241.

- Kiesch, B., "Dating Works of Art through Their Natural Radioactivity: Improvements and Applications," *Science*, 160(April, 1968), 413–415.

Exercises

1. Verify relationship (2.1) between the decay constant k and the half-life of a radioactive material. That is, show that half-life $= \ln(2)/k$.

2. Try to determine if the other two paintings are forgeries.

2.2   Rural Water Supplies

Consider a cylindrical water tank discharging into a pipe in the middle of its base. The potential energy given by the pressure of the column of liquid is converted into kinetic energy of the stream as it exits. If the level of the water in the tank is h ft, then it follows that the exit velocity in ft/sec is $v = \sqrt{2gh}$, where $g = 32.2$ ft/sec^2 is the acceleration of gravity. Since a convergent flow is set up from all sides to the orifice, inertia from the particles in the jet forces them to overshoot the edge and to converge to a smaller cross section termed the *vena contracta*. The ratio of the diameter of the *vena contracta* to the diameter of the orifice is called the contraction coefficient α; and it ranges from 0.5 to 1.0, depending on the smoothness of the edge of the orifice. The diameter of the *vena* cross section and the exit velocity of the stream determine the volume rate of discharge of the stream. Thus, if the tank has cross section A and is being filled from the top at rate K, while draining from the bottom through an orifice of area a, the height in the tank is governed by the differential equation

$$\frac{dh}{dt} = \frac{(K - \alpha a \sqrt{2gh})}{A}.$$

Now transport this first order nonlinear differential equation from the classroom to a summer cabin in the mountains outside Stowe, VT. The water for the cabin is supplied from an aquifer some 125 ft below the surface at a maximum rate of 7 gallons per minute (gpm). Since this flow rate is insufficient to meet daily needs, you must design a more complicated supply system. In your design, this water is pumped to a cylindrical steel tank with diameter 10 ft and height 35 ft which is located on a nearby rise about 75 ft from the cabin, with base 25 ft above the level of the cabin floor. The water is stored in this gravity tank until needed.

Assume that the coefficient of contraction at the exit from the tank is $\alpha = 0.63$, and that the water main consists of 6 inch diameter PVC pipe, so that the friction losses are less than 2 psi (a larger diameter pipe offers less resistance).

a. Determine the minimum flow rate in the system and the minimum water pressure during the first two hours.

b. The local fire code demands that water supplies maintain a pressure of at least 20 psi at a flow rate of 30 gpm for a minimum of two hours.

   Will your water supply design pass inspection?

c. At continuous maximum flow, how long would it take for the level of the top of the water in the tank to fall to 10 ft above the base?

```
>  restart:
>  diffeq:=diff(h(t),t)=(K-alpha*a*sqrt(2.0*g*h(t)))/A;
```

$$diffeq := \frac{\partial}{\partial t}\, h(t) = \frac{K - 1.414213562\, \alpha a\, \sqrt{g\, h(t)}}{A}$$

```
>  inits:=h(0)=35;
```
$$inits := \mathrm{h}(0) = 35$$

```
>  alpha:=0.63;a:=evalf(Pi)*(3/12)^2;A:=evalf(Pi)*5^2;g:=32.2;
```
$$\alpha := .63$$
$$a := .1963495409$$
$$A := 78.53981635$$
$$g := 32.2$$

There are 231 in^3 per gallon.

```
>  K:=evalf(7*(231/12^3));
```
$$K := .9357638889$$

The assignment statements give numerical values in the differential equation.

```
>  diffeq;
```
$$\frac{\partial}{\partial t}\,\mathrm{h}(t) = .01191451588 - .01263931368\,\sqrt{\mathrm{h}(t)}$$

```
>  dsolve({diffeq,inits},h(t));
```

In Release 4, Maple can solve differential equations with all coefficients written as floating point decimals; however here the differential equation is simply too difficult. Since we get no answer, we resort to the `numeric` option.

```
>  sol:=dsolve({diffeq,inits},h(t),numeric);
```
$$sol := \mathbf{proc}(rkf45\_x)\;\ldots\;\mathbf{end}$$

```
>  with(plots):
>  odeplot(sol,[t,h(t)],0..120);
```

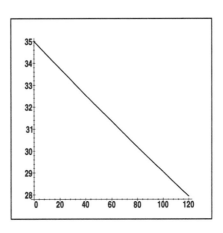

Since the slope remains roughly constant, we check that the fire regulations are met at the two hour mark. The flow rate is estimated by subtracting the heights at the 121st and 120th minutes and multiplying by the area of the base of the tank.

```
>  sol(120);sol(121);
```

$$[t = 120, \text{h}(t) = 27.93672965500731]$$

$$[t = 121, \text{h}(t) = 27.88187162900515]$$

- Since the solution is given as a Maple list, we cannot just subtract `sol(120)-sol(121)`. Instead, we have to extract the second piece of `sol(120)`, which is an equation; and then we extract the second piece of that equation, which is the value of h at $t = 120$. Since it is annoying to have to do this by hand, we write a Maple arrow-defined function H to do it for us. The subscript 2 picks off the second entry in the list; and we use the right hand side of that equation.

```
>  H:=t->rhs(sol(t)[2]);
```

$$H := t \to \text{rhs}(\text{sol}(t)_2)$$

```
>  (H(120)-H(121))*evalf(Pi)*25;
```

$$4.308539603$$

```
>  "*(12^3/231);
```

$$32.23011443$$

Thus, the flow rate at the two hour mark is over 30 gpm. To check the pressure at the orifice, we use the fact that a one foot head of water is equivalent to 0.434 psi.

```
>  (H(120)+25)*0.434;
```

$$22.97454067$$

Since the drop in pressure due to friction in the lines is only 2 psi, the design passes inspection.

- To determine the time to drain the tank to within 10 ft of the bottom, we use `fsolve`. Since `sol` is a numeric procedure, it makes no sense to evaluate it at a variable t. Maple complains that it `cannot evaluate boolean` if one enters the command `sol(t);`. It makes a similar complaint if one uses the command `fsolve(H(t)=35,t);`. One can avoid premature evaluation by means of the same delayed evaluation single quotes that are used to unassign a variable.

```
>  fsolve('H(t)'=10.0,t);
```

$$\text{fsolve}(\text{H}(t) = 10.0, t)$$

Since `fsolve` cannot find an answer on its own, we help it by estimating an interval in which the solution might lie.

```
>  odeplot(sol,[t,h(t)],0..600);
```

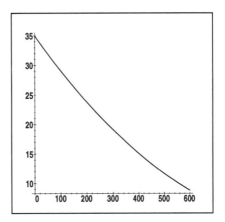

The height drops to 10 ft after an interval somewhere between 500 and 600 minutes. We now include this data in the `fsolve` command.

```
>  fsolve('H(t)'=10.0,t,t=500..600);
```
$$556.0918388$$

Thus, the tank reaches the 10 ft level after 9 hours and 16 minutes.

```
>  unassign('alpha','a','A','g'):
```

For further information, see:

- Campbell, Stu. *The Home Water Supply.* Charlotte, VT: Garden Way Publishing, 1983.

- Hackelman, Michael A. *Waterworks: An Owner-Builder Guide to Rural Water Systems.* Garden City, NY: Dolphin Books, 1983.

Exercises

1. The initial differential equation for $h(t)$ had floating point coefficients and `dsolve` would not solve it. In fact, even after we had converted all coefficients to fractions with integer numerators and denominators, `dsolve` was still unable to give an answer. Use `dsolve` just on the differential equation, i.e., leave off the initial conditions; and verify that the real difficulty is not in finding a general solution, but instead is in solving for the constants of integration that match the initial conditions.

2. Consider the solution $H(t)$ in Section 2.2 that gives the height of the water level in the tank, relative to its base. Calculate $H(1000)$, $H(2000)$, $H(3000)$, $H(4000)$, and $H(5000)$. Now compare these values with the equilibrium solution. Since computing values with numeric methods for values of t which

are large will usually yield large errors, are you surprised by the accuracy? Plot the direction field and discuss how it relates to this amazing accuracy.

2.3 Growth of Microorganisms: Linear and Nonlinear

One of the simplest biological experiments is growing bacteria in a petri dish and following the change in the bacterial population over time.[1]

Important considerations are the size of the dish, the amount of nutrient (referred to as the *medium*) necessary to feed the bacteria in the dish, and environmental conditions (such as temperature, light, etc.). In such situations, what is typically seen is that the bacteria grow exponentially for at least short periods of time. There are two main approaches to this type of problem: using *linear* first order differential equations and using *nonlinear* ones. In this example we use both methods in order to compare them.

Define $P(t)$ to be the bacterial population at time t. Suppose we are able to observe, over a period of one unit time, as a single bacterial cell divides, its daughters divide, and so forth, leading to a total of r new bacteria. Define r to be the rate of reproduction per unit time. The governing differential equation, assuming the law of exponential growth, is:

$$\frac{dP}{dt} = rP, \ P(0) = P_0,$$

where $P(0) = P_0$ is the *initial condition*: at time zero, the start of the experiment, we start with P_0 bacteria.

We solve this *linear* first order differential equation as follows:

```
>  restart:
```

```
>  diffeq1:=diff(P(t),t)=r*P(t);inits:=P(0)=P[0]:
```

$$diffeq1 := \frac{\partial}{\partial t} P(t) = r P(t)$$

```
>  P=rhs(dsolve({diffeq1,inits},P(t)));
```

$$P = e^{(r\,t)} P_0$$

Let's examine the numerical output and look at the solutions. To do this, we need to have a growth rate r and different initial population sizes P_0. For the bacterium *Helicobacter pylori*, it is known that the time it takes for some initial population size, P_0, to double (called the *doubling time*) is 7 days. Suppose we begin with 100,000 bacteria in a petri dish. This means it will take 7 days for the population to increase in size to 200,000. From this information we can find the growth rate, r. Examine the solution to *eq1* above. We can put in the known information as follows and find r.

```
>  eq2:=200000=100000*exp(7*r);
```

$$eq2 := 200000 = 100000\, e^{(7\,r)}$$

```
>  r:=solve(eq2,r);
```

$$r := \frac{1}{7} \ln(2)$$

---

[1]The authors would like to thank Professor Denise E. Kirschner of the Texas A&M University Department of Mathematics for her suggestions on this section.

Now, we have calculated the growth rate r. Rewriting the solution to *eq2* to describe this particular example, we have:

```
>   P:=100000*exp(r*t);
```

$$P := 100000\, e^{(1/7\ln(2)\, t)}$$

Now we can plot the solution and see the results numerically. Let's run the experiment for 30 days.

```
>   with(plots):
```

```
>   p1:=plot(P,t=0..30):
```

```
>   plot(P,t=0..30);
```

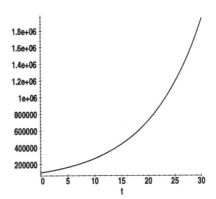

We can compare this result with different initial conditions to see how it changes the result. Suppose we start with 50,000 bacteria instead of 100,000. How does this change the problem? Does the doubling time change? Does the growth rate r change? No, the only thing that changes is the initial condition.

```
>   P:=50000*exp(r*t);
```

$$P := 50000\, e^{(1/7\ln(2)\, t)}$$

```
>   p2:=plot(P,t=0..30):
```

```
>   display([p1,p2]);
```

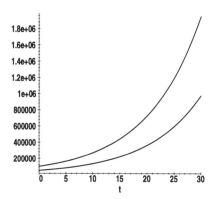

It is clear that no matter what initial population size we start with, the population of bacteria in each case grows to infinity. This constitutes the simplest model of bacterial growth. In 1798 it was first applied by Thomas Robert Malthus (1766–1834) to human populations, and his conclusions worried many people. He claimed that barring natural disasters, man would eventually outgrow the world's resources and we would starve to death. The question is, then, what is wrong with this model?

We have left out some very important assumptions which may be biologically significant. Here are a few:

1. The food supply limits growth. This means that if there is some initial amount of medium, eventually it will be eaten; and the population of bacteria will not only stop growing, it will most likely decrease.

2. Are we sure the growth rate r is constant? Perhaps under different environmental or biological conditions, the growth rate is actually time dependent. Another possibility is that the growth rate depends on the amount of nutrient available to the population.

Let's work with this second idea, in hope of also satisfying the first. Define $N(t)$ to be the nutrient population at time t. Perhaps the growth rate is now a function of the nutrient, i.e. $r = r(N)$. Let's assume the simplest relationship, in other words that the growth rate is directly proportional to the nutrient, with proportionality constant r_1. Therefore,

$$r(N) = r_1 N. \tag{2.2}$$

The resulting differential equation is

```
>  diffeq3:=diff(P1(t),t)=r1*N(t)*P1(t);
```

$$diffeq3 := \frac{\partial}{\partial t} P1(t) = r1\, N(t)\, P1(t)$$

There is one more issue to be resolved before we proceed. Do we know for certain that the relationship between bacteria production and nutrient consumption is one to one? In other words, does one unit of nutrient produce one bacterium? To handle this idea, we introduce a new parameter $Y = 1/a$, called the *growth yield constant*. In this case, a units of nutrient are needed to produce one unit of bacteria. The rate of change in the amount of nutrient is proportional to the change in the bacterial population via the proportionality constant a.

We can write this as

$$diffeq4 := \frac{\partial}{\partial t} N(t) = -a\, r1\, P1(t)\, N(t).$$

This system of two differential equations can be turned into a single differential equation by substituting *diffeq3* into *diffeq4*. This gives the new equation:

$$diffeq5 := \frac{\partial}{\partial t} N(t) = -a \left(\frac{\partial}{\partial t} P1(t) \right).$$

We can now integrate.

```
>  eq6:=N(t)=-P1(t)*a+P1(0)*a+N[0];
```

$$eq6 := N(t) = -P1(t)\, a + P1(0)\, a + N_0$$

Notice that the last two terms of this equation are *both* constants. We can incorporate both terms in a constant $\hat{N}_0$. Therefore, the solution is of the form

$$N(t) = -aP1(t) + \hat{N}_0.$$

Now, substitute $N(t)$ from *eq6* into *diffeq3* and we have:

```
>  diffeq7:=diff(P1(t),t)=r1*(N[0]-a*P1(t))*P1(t);
```

$$diffeq7 := \frac{\partial}{\partial t} P1(t) = r1\, (N_0 - P1(t)\, a)\, P1(t)$$

Notice that we now have a *nonlinear* first order differential equation. We can again use `dsolve`. The solution to this equation is:

```
>  eq8:=P1=rhs(dsolve({diffeq7,P1(0)=P1[0]},P1(t))):
>  simplify(");
```

$$P1 = \frac{P1_0\, N_0}{P1_0\, a + e^{(-r1\, N_0\, t)}\, N_0 - e^{(-r1\, N_0\, t)}\, P1_0\, a}$$

Look how much more complicated the solution is, now that we are solving a nonlinear differential equation. We can approximate the solution numerically to compare the differences between the *linear* and the *nonlinear* cases. We must choose different values for the growth rate r and different initial population sizes $P1_0$; and we must also choose the medium nutrient amount N_0 as well. Let's examine the same situation as above with the *Helicobacter pylori* bacteria population. We already know the growth rate r. Let us also consider identical initial conditions so that we can compare our results. The new information we require is the nutrient size and the growth yield constant a. To simplify things, we will choose $a = 1$. It remains to choose the medium size. Let's begin with $N_0 = 2,000,000$. Thus *eq8*, with numerical values for r and P_0, is:

```
>  P1[0]:=100000;N[0]:=2000000;r1:=r;a:=1;
```

$$P1_0 := 100000$$

$$N_0 := 2000000$$

$$r1 := \frac{1}{7} \ln(2)$$

$$a := 1$$

We can plot the solution and see the results below.

```
>  P1:=rhs(eq8):
```

```
>  p3:=plot(P1,t=0..1e-04,tickmarks=[5,5]):
```

We can compare this result with that obtained from different initial conditions to see how it changes the result. Suppose we start with 5,000,000 bacteria instead of 100,000.

```
>  P1[0]:=5000000;
```

$$P1_0 := 5000000$$

```
>  P1:=rhs(eq8);
```

$$P1 := \frac{2000000}{1 - \dfrac{3}{5} e^{(-2000000/7 \ln(2)\, t)}}$$

```
>  p4:=plot(P1,t=0..1e-04,tickmarks=[5,5]):
```

```
>  display([p3,p4],tickmarks=[5,5]);
```

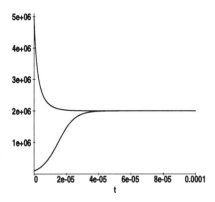

The populations level off to the value of $N_0/a = 2,000,000$. Notice that the trajectory which starts at the initial condition less than 2,000,000 approaches it from below, while the trajectory which begins at the initial condition greater than 2,000,000 approaches it from above. We refer to this value as the *carrying capacity* of the problem. The unrealistic result of growth to infinity, as seen in the linear case above, is now replaced by an asymptotic leveling off to the value of the carrying capacity. This scenario makes better sense in the study of population growth. For further information, see:

1. Edelstein-Keshet, L. *Mathematical Models in Biology*. New York: Random House, Inc., 1988.

2. Giordano, F. and M. Weir. *A First Course in Mathematical Modeling*. Pacific Grove, CA: Brooks/Cole Publishing Co., 1985.

Exercises

1. How do the solutions vary as a function of the doubling time?

   (a) Consider a population of bacteria which grows at a much faster rate than *Helicobacter pylori*. For example, *Escherichia coli* has a doubling time of 3.4 days. How does this change the solutions in both the linear and nonlinear cases?

   (b) Consider a population which grows at a much slower rate than *H. pylori*. For example, *H. sapiens* has a doubling time of (approximately) 50 years (scary!). How does this change the behavior of solutions in both the linear and nonlinear cases?

2. In the calculations for the nonlinear case, we assumed the simplest scenario: $r(N) = r_1 N$. What if the relationship is not a direct proportion, but some other functional form? For example, rework the calculations with $r(N) = r_1 N^2$ and see how that modification changes things in the nonlinear case.

Chapter 3

Higher Order Equations

In this chapter we will use Maple to solve second and higher order differential equations. In some simple situations the Maple command dsolve gives satisfactory results; in other situations Maple may give a correct answer, but in an awkward form. For this reasons we will concentrate on using Maple along with the techniques you learned in class to solve higher order equations.

3.1 dsolve, dsolve/numeric, and DEplot

Second and higher order equations are more complicated than first order equations, but often they may be solved directly or approximated numerically with the same Maple tools that we used for first order problems, using essentially the same syntax.

Consider the second order initial value problem given by $y'' + 2y' + y = x^2 + 1 + e^x$, $y(0) = 0$, $y'(0) = 2$.

```
>  restart:
>  diffeq1:=diff(y(x),x$2)+2*diff(y(x),x)+y(x)=x^2+1+exp(x);
```
$$diffeq1 := (\frac{\partial^2}{\partial x^2} y(x)) + 2 (\frac{\partial}{\partial x} y(x)) + y(x) = x^2 + 1 + e^x$$

```
>  inits1:=y(0)=0,D(y)(0)=2;
```
$$inits1 := y(0) = 0, \ D(y)(0) = 2$$

To get the general solution to the differential equation, we use:

```
>  sol1g:=dsolve(diffeq1,y(x));
```
$$sol1g := y(x) = x^2 - 4x + 7 + \frac{1}{4} e^x + \_C1 \, e^{(-x)} + \_C2 \, e^{(-x)} x$$

To solve the initial value problem, we use:

```
>  dsolve({diffeq1,inits1},y(x));
```
$$y(x) = \frac{1}{4} \frac{4x^2 e^x - 16x e^x + 28 e^x + (e^x)^2 - 29 - 6x}{e^x}$$

If you prefer a different form for the answer, try:

```
> expand(");sol1p:=simplify(",exp);
```

$$y(x) = x^2 - 4x + 7 + \frac{1}{4}e^x - \frac{29}{4}\frac{1}{e^x} - \frac{3}{2}\frac{x}{e^x}$$

$$sol1p := y(x) = x^2 - 4x + 7 + \frac{1}{4}e^x - \frac{29}{4}e^{(-x)} - \frac{3}{2}e^{(-x)}x$$

To plot the solution of the initial value problem on the interval $[-1, 3]$, use:

```
> plot(rhs(sol1p),x=-1..3);
```

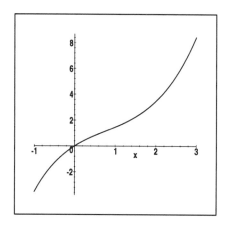

To obtain a floating point decimal value for the solution at $x = 2.5$, use:

```
> subs(x=2.5,sol1p);
```

$$y(2.5) = 3.25 + \frac{1}{4}e^{2.5} - 11.00000000\,e^{(-2.5)}$$

```
> evalf(");
```

$$y(2.5) = 5.392688505$$

For a third order initial value problem, $y''' + y'' + 3y' - 5y = 2 + 6x - 5x^2$, $y(0) = -1$, $y'(0) = 1$, $y''(0) = -3$, we merely include the second derivative entry in the initial value sequence:

```
> diffeq2:=diff(y(x),x$3)+diff(y(x),x$2)+3*diff(y(x),x)-5*y(x)=2+6*x-5*x^2;
```

$$diffeq2 := (\frac{\partial^3}{\partial x^3}y(x)) + (\frac{\partial^2}{\partial x^2}y(x)) + 3\,(\frac{\partial}{\partial x}y(x)) - 5\,y(x) = 2 + 6\,x - 5\,x^2$$

```
> inits2:=y(0)=-1,D(y)(0)=1,(D@@2)(y)(0)=-3;
```

$$inits2 := y(0) = -1,\; D(y)(0) = 1,\; (D^{(2)})(y)(0) = -3$$

```
>  dsolve({diffeq2,inits2},y(x));
```

$$y(x) = \frac{x^2\,e^x - (e^x)^2 + \sin(2\,x)}{e^x}$$

```
>  expand(");sol2:=simplify(",exp);
```

$$y(x) = x^2 - e^x + 2\,\frac{\sin(x)\cos(x)}{e^x}$$

$$sol2 := y(x) = x^2 - e^x + 2\,e^{(-x)}\sin(x)\cos(x)$$

It is also possible to approximate numerically a second order equation solution that dsolve cannot find. Consider the nonlinear initial value problem with *numerical initial values*: $t^2 y'' - y y' = y^{-3}$, $y(1) = 0.5$, $y'(1) = 2.1$.

```
>  diffeq3:=t^2*diff(y(t),t$2)-y(t)*diff(y(t),t)=y(t)^(-3);
```

$$diffeq3 := t^2\,(\frac{\partial^2}{\partial t^2}\,y(t)) - y(t)\,(\frac{\partial}{\partial t}\,y(t)) = \frac{1}{y(t)^3}$$

```
>  inits3:=y(1)=0.5,D(y)(1)=2.1;
```

$$inits3 := y(1) = .5,\ D(y)(1) = 2.1$$

```
>  sol3:=dsolve({diffeq3,inits3},y(t),numeric);
```

$$sol3 := \mathbf{proc}(rkf45\_x)\ \dots\ \mathbf{end}$$

The approximate values of the solution and of the derivative of the solution at x=1.5 are given by:

```
>  sol3(1.5);
```

$$[t = 1.5,\ y(t) = 2.044481198876178,\ \frac{\partial}{\partial t}\,y(t) = 3.873909130886410]$$

The easiest way to plot a numerical solution on the interval $[0.6, 1.2]$ is to use plots[odeplot].

```
>  with(plots):
```

```
>  odeplot(sol3,[t,y(t)],0.6..1.2);
```

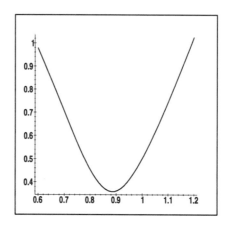

A common difficulty is failure to write the `odeplot` command with the same variables in it that were in the problem when we called `dsolve/numeric`. Axes will then be shown, but no graph will be produced.

```
>  odeplot(sol3,[x,y(x)],0.6..1.2);
```

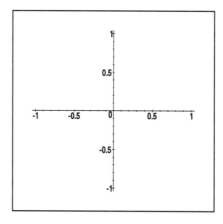

Although there is no way to draw direction field arrows with a second order differential equation, we can still use DEplot to approximate solutions using the Runge-Kutta method. Note that the initial conditions *cannot* be indicated as lists of lists of numbers, the way that they were with first order equations.

```
>  with(DEtools):
```

```
>  DEplot(diffeq3,y(t),t=0.6..1.2,[[y(1)=0.5,D(y)(1)=2.1]]);
```

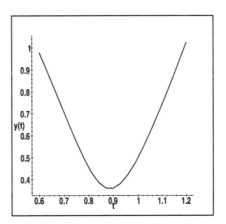

In the following sections we will study the classical solution methods, with Maple doing the arithmetic for us at intermediate steps, rather than having Maple run on autopilot. Situations can arise where we will be able to assist Maple to find a solution that it cannot find on its own, or where we will be able to guide Maple to find nicer solutions than it would find on its own.

3.2 Homogeneous equations

As we have just seen, Maple's `dsolve` command is quite adequate for solving second order homogeneous equations. As another example, let us find the general solution of

$$y'' + y' + y = 0.$$

```
> restart:
> diffeq:=diff(y(t),t$2)+diff(y(t),t)+y(t)=0;
```

$$diffeq := (\frac{\partial^2}{\partial t^2} \, y(t)) + (\frac{\partial}{\partial t} \, y(t)) + y(t) = 0$$

```
> sol:=dsolve(",y(t));
```

$$sol := y(t) = \_C1 \, e^{(-1/2\,t)} \sin(\frac{1}{2} \, \sqrt{3}\,t) + \_C2 \, e^{(-1/2\,t)} \cos(\frac{1}{2} \, \sqrt{3}\,t)$$

The individual functions which comprise the fundamental set of solutions to this equation can be obtained by

```
> y1:=rhs(subs({_C2=0,_C1=1},sol));
```

$$y1 := e^{(-1/2\,t)} \sin(\frac{1}{2} \, \sqrt{3}\,t)$$

and

```
> y2:=rhs(subs({_C2=1,_C1=0},sol));
```

$$y2 := e^{(-1/2\,t)} \cos(\frac{1}{2}\sqrt{3}\,t)$$

For higher order equations Maple's `dsolve` is not adequate. If we use it to solve

$$y^{(4)} + y' + y = 0,$$

we get

```
> diffeq:=diff(y(t),t$4)+diff(y(t),t)+y(t)=0;
```

$$diffeq := (\frac{\partial^4}{\partial t^4}\,y(t)) + (\frac{\partial}{\partial t}\,y(t)) + y(t) = 0$$

```
> sol:=dsolve(",y(t));
```

$$sol := y(t) = \sum_{\_R=RootOf(1+\_Z+\_Z^4)} \_C1_{\_R}\,e^{(-R\,t)}$$

We can use other Maple commands to get to the solution of this equation. Solutions to this equation have the form $y = e^{rt}$, where r is any root of the characteristic equation $r^4 + r + 1 = 0$.

```
> ev:= fsolve(r^4+r+1=0,r,complex);
```

$$ev := -.7271360845 - .4300142883\,I, \; -.7271360845 + .4300142883\,I,$$
$$.7271360845 - .9340992895\,I, .7271360845 + .9340992895\,I$$

We extract each root, substitute into e^{rt}, and convert to standard complex form. (Note that e^w is expressed as `exp(w)` in Maple.)

```
> Y1:=evalc(exp(ev[1]*t));
```

$$Y1 := e^{(-.7271360845\,t)} \cos(.4300142883\,t) - I\,e^{(-.7271360845\,t)} \sin(.4300142883\,t)$$

```
> Y2:=evalc(exp(ev[2]*t));
```

$$Y2 := e^{(-.7271360845\,t)} \cos(.4300142883\,t) + I\,e^{(-.7271360845\,t)} \sin(.4300142883\,t)$$

```
> Y3:=evalc(exp(ev[3]*t));
```

$$Y3 := e^{(.7271360845\,t)} \cos(.9340992895\,t) - I\,e^{(.7271360845\,t)} \sin(.9340992895\,t)$$

```
> Y4:=evalc(exp(ev[4]*t));
```

$$Y4 := e^{(.7271360845\,t)} \cos(.9340992895\,t) + I\,e^{(.7271360845\,t)} \sin(.9340992895\,t)$$

Finally, we notice that *Y1* and *Y2*, as well as *Y3* and *Y4*, form a conjugate pair. Taking linear combinations of these with *complex* multiples, we finally get the solutions in real form.

```
> y1:=(Y1+Y2)/2;
```

$$y1 := e^{(-.7271360845\,t)} \cos(.4300142883\,t)$$

```
> y2:=(Y2-Y1)/(2*I);
```

$$y2 := e^{(-.7271360845\, t)} \sin(.4300142883\, t)$$

Alternatively, we may use Re and Im to extract the real and imaginary parts of a complex expression, respectively.

```
> y3:=evalc(Re(Y3));
```

$$y3 := e^{(.7271360845\, t)} \cos(.9340992895\, t)$$

```
> y4:=evalc(Im(Y4));
```

$$y4 := e^{(.7271360845\, t)} \sin(.9340992895\, t)$$

A fundamental set of real solutions to our 4th order equation is given by *y1*, *y2*, *y3*, and *y4*.

3.3 Nonhomogeneous Equations

In order to illustrate the usefulness of Maple in solving nonhomogeneous equations, we will solve a fairly simple initial value problem — one that would require lengthy calculations if done by hand. Consider the initial value problem

$$y'' + y = t \cos t, \quad y(0) = 1, \quad y'(0) = 1.$$

This problem will be solved first using the method of undetermined coefficients and second using the method of variation of parameters.

3.3.1 Undetermined coefficients

```
> diffeq:=diff(y(t),t$2)+y(t)=t*cos(t);
```

$$diffeq := (\frac{\partial^2}{\partial t^2}\, y(t)) + y(t) = t \cos(t)$$

Solving the homogeneous equation is quite easy.

```
> hsol:=dsolve(lhs(diffeq),y(t));
```

$$hsol := y(t) = \_C1 \sin(t) + \_C2 \cos(t)$$

Now set up a particular solution, noting that the nonhomogeneous term is a product of a linear polynomial and the cosine function. Our guess for the particular solution must consist of general linear polynomials and sine and cosine functions.

```
> psol:= y(t)=(a*t+b)*sin(t)+(c*t+d)*cos(t);
```

$$psol := y(t) = (a\, t + b) \sin(t) + (c\, t + d) \cos(t)$$

Now substitute *psol* into the differential equation and simplify.

```
> subs(psol,diffeq):
> simplify(");
```

$$2\, a \cos(t) - 2\, c \sin(t) = t \cos(t)$$

Clearly there is no solution for a and c since the left-hand side has no $t \sin t$ term.

Examining our first guess for *psol* we see that *psol* contains terms which are solutions to the homogeneous equation, so we have to modify *psol* by multiplying by the independent variable.

```
>   psol:=y(t)=rhs(psol)*t;
```

$$psol := \mathrm{y}(t) = ((a\,t + b)\sin(t) + (c\,t + d)\cos(t))\,t$$

Substitute our new guess for *psol* into the differential equation and simplify.

```
>   subs(psol,diffeq):
```

```
>   simplify(");
```

$$4\,t\,a\cos(t) - 4\,t\,c\sin(t) + 2\,a\sin(t) + 2\cos(t)\,b + 2\,c\cos(t) - 2\sin(t)\,d = t\cos(t)$$

Maple has a pattern matching capability, which it can use to turn an equation into an `identity`. The variable for which the statement is always supposed to be true is the second argument (there might be other variables in the equation). Once the identity is formed, we can use `solve` to determine the values of a, b, c, and d for which it will remain true for all t.

```
>   identity(",t):
```

```
>   csol:=solve(",{a,b,c,d});
```

$$csol := \{c = 0,\ b = 0,\ d = \frac{1}{4},\ a = \frac{1}{4}\}$$

Substituting the coefficients back into *psol*, we finally have a particular solution.

```
>   subs(csol,psol):
```

```
>   expand(");
```

$$\mathrm{y}(t) = \frac{1}{4}t^2\sin(t) + \frac{1}{4}t\cos(t)$$

Adding the homogeneous solution to our particular solution, we now have our general solution.

```
>   gsol:=y(t)=rhs(hsol)+rhs(");
```

$$gsol := \mathrm{y}(t) = \_C1\sin(t) + \_C2\cos(t) + \frac{1}{4}t^2\sin(t) + \frac{1}{4}t\cos(t)$$

All that remains is to use the initial conditions $y(0) = 1$ and $y'(0) = 1$ to solve for $\_C1$ and $\_C2$.

```
>   subs(t=0,rhs(gsol))=1:
```

```
>   eq1:=simplify(");
```

$$eq1 := \_C2 = 1$$

```
>   diff(rhs(gsol),t):
```

```
>   subs(t=0,")=1:
```

```
>   eq2:=simplify(");
```

$$eq2 := \_C1 + \frac{1}{4} = 1$$

Solve this linear system for _C1 and _C2.

```
> csol:=solve({eq1,eq2},{_C1,_C2});
```

$$csol := \{\_C1 = \frac{3}{4}, \_C2 = 1\}$$

Putting it all together we finally have the solution to the initial value problem.

```
> subs(csol,gsol);
```

$$y(t) = \frac{3}{4}\sin(t) + \cos(t) + \frac{1}{4}t^2\sin(t) + \frac{1}{4}t\cos(t)$$

3.3.2   Variation of Parameters

When the right hand side of a differential equation is of a form that we can use the method of undetermined coefficients, we generally prefer to do so. However, it is often the case that the right hand side cannot be changed to a form which is amenable to the undetermined coefficients approach; and we then use the method of variation of parameters, or *variation of constants*, as it used to be known.

We will demonstrate the method on the differential equation that we just solved by the method of undetermined coefficients. As with that method, we must first find the solutions of the homogeneous equation.

```
> hsol:=dsolve(lhs(diffeq),y(t));
```

$$hsol := y(t) = \_C1 \sin(t) + \_C2 \cos(t)$$

The idea of variation of parameters is to replace the constants in the solution to the homogeneous equation by functions, i.e., to *vary the constants* with u_1 and u_2, where u_1 and u_2 are to be solved for.

```
> psol:=subs({_C1=u1(t),_C2=u2(t)},hsol);
```

$$psol := y(t) = u1(t)\sin(t) + u2(t)\cos(t)$$

The derivatives u_1' and u_2' satisfy two equations:

$$u_1'y_1 + u_2'y_2 = 0,$$
$$u_1'y_1' + u_2'y_2' = t\cos(t).$$

We denote u_1' and u_2' by *u1p* and *u2p*, respectively; and we set up the equations.

```
> y1:=cos(t):y2:=sin(t):
```

```
> eq1:=u1p*y1+u2p*y2=0;
```

$$eq1 := u1p\cos(t) + u2p\sin(t) = 0$$

```
> eq2:=u1p*diff(y1,t)+u2p*diff(y2,t)=t*cos(t);
```

$$eq2 := -u1p\sin(t) + u2p\cos(t) = t\cos(t)$$

We then ask Maple to solve:

```
> solve({eq1,eq2},{u1p,u2p}):
```

```
> convert(",sincos):
```

```
> upsol:=simplify(");
```

$$upsol := \{u1p = -t\cos(t)\sin(t),\ u2p = t\cos(t)^2\}$$

We need to integrate to find $u_1(t)$ and $u_2(t)$.

```
>   subs(upsol,u1p):
>   u1(t):=int(",t);
```

$$u1(t) := \frac{1}{2} t \cos(t)^2 - \frac{1}{4} \cos(t) \sin(t) - \frac{1}{4} t$$

```
>   subs(upsol,u2p):
>   int(",t):
>   u2(t):=simplify(");
```

$$u2(t) := \frac{1}{2} t \cos(t) \sin(t) + \frac{1}{4} t^2 + \frac{1}{4} \cos(t)^2$$

We can then get the particular solution:

```
>   y(t)=rhs(psol):
>   psol:=simplify(");
```

$$psol := \mathrm{y}(t) = \sin(t) t \cos(t)^2 - \frac{1}{4} \sin(t) t + \frac{1}{4} \cos(t) t^2 + \frac{1}{2} \cos(t)^3 - \frac{1}{4} \cos(t)$$

Adding the homogeneous solution, we get the general solution:

```
>   rhs(psol)+rhs(hsol):
>   gsol:=y(t)=simplify(");
```

$$gsol := \mathrm{y}(t) =$$
$$\sin(t) t \cos(t)^2 - \frac{1}{4} \sin(t) t + \frac{1}{4} \cos(t) t^2 + \frac{1}{2} \cos(t)^3 - \frac{1}{4} \cos(t) + \_C1 \sin(t) + \_C2 \cos(t)$$

Note that the solution agrees with the earlier result. We could then continue to solve for the constants of integration as before.

We can also use the built-in Release 4 Maple command, which uses the fundamental solutions and the right hand side of the differential equation (written in *standard form*) as its inputs.

```
>   y(t)=DEtools[varparam]([sin(t),cos(t)],rhs(diffeq),t):
>   simplify(");
```

$$\mathrm{y}(t) = \_C_1 \sin(t) + \_C_2 \cos(t) + \frac{1}{4} t \cos(t) + \frac{1}{4} t^2 \sin(t)$$

3.4   Exercises

Use the methods of undetermined coefficients and variation of parameters to find the general solution to each of the following equations.

1. $y'' + 4y' + 8y = 8t^2 + 16t^3 - 12t^2 - 24t - 6$

2. $y'' + 4y' + 5y = 60e^{-2t} \sin t$

3. $y'' - y' - 2y = 4t^2 - \sin 2t$

4. $y'' + 5y' + 6y = 3e^{-2t} + e^{3t}$

5. $y'' + 4y = \sin^2(2t)$

6. $y'' - 4y' + 3y = \dfrac{1}{1 + e^t}$

7. $y'' - y = e^{-t}\sin(e^{-t}) + \cos(e^{-t})$

8. $y'' - 2y' = e^t \sin t$

9. Consider the nonhomogeneous differential equation $y''' - 2y'' - y' + 2y = e^{4x}$ in standard form. Solve it by variation of parameters in the following manner.

   (a) Use Maple to solve the characteristic equation and make a list of independent solutions to the homogeneous problem.

   (b) Convert the list to a vector, **isols.**

   (c) Use `with(linalg):W:=Wronskian(isols,x);` to form the Wronskian matrix for the set of independent solutions.

   (d) Form a 3×1 column matrix **G** with zeros in each entry. Then assign `G[3,1]:=exp(4*x);` i.e., let it be the term on the right side that makes the equation nonhomogeneous.

   (e) Solve for the matrix **uprimes** by `uprimes:=(1/W) &* G;`.

   (f) Use the `map` command to integrate **uprimes**, convert the result to a vector **u**, and find a particular solution to the nonhomogeneous equation by taking the dot product with **isols**.

10. Given a homogeneous differential equation $y'' + py' + qy = 0$, and a nontrivial solution $y = y_1(x)$, one can find a second linearly independent solution by reduction of order. For example, given the equation $y'' + 10y' + 25y = 0$, we know from the characteristic equation that one solution is $y(x) = e^{-5x}$. Use the Release 4 command `reduceOrder` in the DEtools package to find a second solution:

```
> diffeq:=diff(y(x),x$2)+10*diff(y(x),x)+25*y(x)=0;
> sol:=exp(-5*x);
> with(DEtools):
> reduceOrder(diffeq,y(x),sol,basis);
```

Chapter 4

Applications of Higher Order Equations

This chapter will be concerned with the application of second order linear differential equations to the study of mechanical vibrations. Such differential equations have the form

$$mu''(t) + cu'(t) + ku(t) = f(t),$$

where m is the mass of the object attached to the spring, c is the damping constant, k is the spring constant and $f(t)$ is an applied force acting on the spring-mass system.

4.1 Beats and Resonance

First we examine the phenomena of beats and resonance. They occur when the damping constant c is zero and $f(t)$ is a periodic applied force. Our model equation becomes

$$mu'' + ku = F \cos{(wt)}.$$

We assume that the mass is initially at rest at the equilibrium position so that $u(0) = 0$ and $u'(0) = 0$. Your textbook contains an analysis of spring-mass systems and we will not be concerned with the general problem here. Rather we will look at some specific examples and use Maple to solve them. For simplicity we will take $m = 1$ and $k = 1$, so that the natural frequency of our spring-mass system is 1. We examine solutions of

$$u'' + u = 2 \cos{(wt)}, \quad u(0) = 0, \quad u'(0) = 0$$

for various values of w.

4.1.1 Beats

Beats occur when $w \neq 1$, so that the frequency of the applied force is different than the natural frequency of the spring-mass system.

We enter the differential equations corresponding to the homogeneous and nonhomogeneous cases, respectively.

$$
\begin{aligned}
u'' + u &= 0 \\
u'' + u &= 2 \cos{(wt)}.
\end{aligned}
$$

```
> diffeq1:=diff(u(t),t$2)+u(t)=2*cos(w*t);
```

$$diffeq1 := (\frac{\partial^2}{\partial t^2} u(t)) + u(t) = 2\cos(w\,t)$$

```
> inits:=u(0)=0,D(u)(0)=0;
```

$$inits := u(0) = 0,\ D(u)(0) = 0$$

When Maple is left to its own devices, it chooses to solve the following equation by the method of undetermined coefficients. The answer that it gives is correct, but suffers from a lack of trigonometric simplification. Instead, we force dsolve to solve the equation using the method of Laplace transforms so that we see the solution more clearly.

```
> sol:=dsolve({diffeq1,inits},u(t),laplace);
```

$$sol := u(t) = 2\,\frac{\cos(t)}{(w-1)\,(w+1)} - 2\,\frac{\cos(w\,t)}{(w-1)\,(w+1)}$$

```
> combine(");
```

$$u(t) = \frac{2\cos(t) - 2\cos(w\,t)}{w^2 - 1}$$

The numerator can be viewed as a product of a more rapidly oscillating function $\sin((w+1)t/2)$ and a more slowly varying function $\sin((1-w)t/2)$:

```
> sin((1/2)*(1-w)*t)*sin((1/2)*(1+w)*t):
```

```
> "=combine(",trig);
```

$$\sin(\frac{1}{2}(1-w)\,t)\sin(\frac{1}{2}(w+1)\,t) = -\frac{1}{2}\cos(t) + \frac{1}{2}\cos(w\,t)$$

To illustrate this behavior with a few graphs, let's plot the resulting solutions for $w = 0.7, 0.8, 0.9$, and 0.95.

```
> T:=title='Beats:  w=0.7':
```

```
> plot(subs(w=0.7,rhs(sol)),t=0..160,T);
```

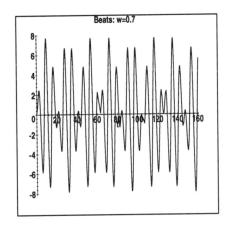

```
> T:=title='Beats:   w=0.8':
```

```
> plot(subs(w=0.8,rhs(sol)),t=0..160,T);
```

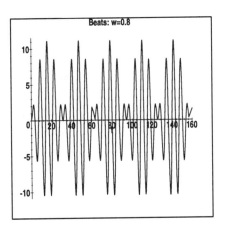

```
> T:=title='Beats:   w=0.9':
```

```
> plot(subs(w=0.9,rhs(sol)),t=0..160,T);
```

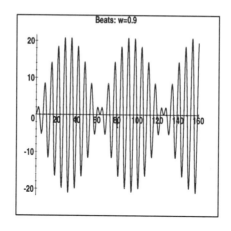

```
>  T:=title='Beats:  w=0.95':
```

```
>  plot(subs(w=0.95,rhs(sol)),t=0..160,T);
```

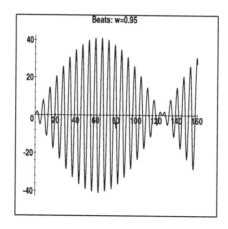

In the above discussion we have been very careful not to let $w = 1$. When the frequency w of the applied force matches the natural frequency of the spring-mass system, we no longer have beats. Instead we encounter the phenomenon of resonance.

4.1.2 Resonance

To illustrate resonance we take $w = 1$, so that our differential equation becomes

$$u'' + u = 2\cos(t), \quad u(0) = 0, \quad u'(0) = 0.$$

```
>  diffeq2:=diff(u(t),t$2)+u(t)=2*cos(t);
```

$$diffeq2 := (\frac{\partial^2}{\partial t^2}\,\mathrm{u}(t)) + \mathrm{u}(t) = 2\cos(t)$$

```
>   inits:=u(0)=0,D(u)(0)=0;
```
$$inits := \mathrm{u}(0) = 0,\ \mathrm{D}(u)(0) = 0$$

We use `dsolve`.

```
>   sol2:=dsolve({diffeq2,inits},u(t));
```
$$sol2 := \mathrm{u}(t) = \sin(t)\,t$$

To see the contrast between resonance and beats, plot the solution we have just found

```
>   T:=title='Resonance':
```

```
>   plot(rhs(sol2),t=0..160,T);
```

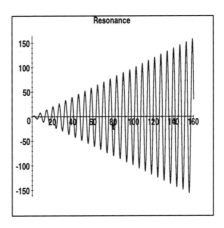

The amplitude of the solution becomes unbounded as t gets large.

4.2   Damped Forced Vibrations

For our last application we will examine the full spring-mass system with a periodic applied force. We will assume that $m = 1$, $c = 1/8$, and $k = 1$. Our goal is to determine the frequency w that gives the largest amplitude. Our initial value problem becomes

$$u'' + \frac{1}{8}u' + u = 2\cos(wt), \quad u(0) = u0, \quad u'(0) = u1.$$

```
>   diffeq3:=diff(u(t),t$2)+(1/8)*diff(u(t),t)+u(t)=2*cos(w*t);
```
$$diffeq3 := (\frac{\partial^2}{\partial t^2}\,\mathrm{u}(t)) + \frac{1}{8}\,(\frac{\partial}{\partial t}\,\mathrm{u}(t)) + \mathrm{u}(t) = 2\cos(w\,t)$$

```
> inits:=u(0)=u0,D(u)(0)=u1;
```

$$inits := u(0) = u0, \ D(u)(0) = u1$$

The solution is very complicated, consisting of two parts: a transient response containing $e^{-t/16}$; and a steady state component, which does not contain it. We can let Maple select the steady state component for us. (Since the answer is so long, Maple tries to abbreviate it by labelling with %1, %2, and %3. We can stop it from doing so by turning off the labelling. When using the `select` and `remove` commands, always look carefully to verify that Maple is finding and choosing the correct terms!)

```
> interface(labelling=false):
```

```
> sol3:=dsolve({diffeq3,inits},u(t),laplace):
```

```
> steady:=remove(has,rhs("),exp);
```

$$steady :=$$
$$-128 \, \frac{w^2 \cos(w \, t)}{64 \, w^4 + 64 - 127 \, w^2} + 128 \, \frac{\cos(w \, t)}{64 \, w^4 + 64 - 127 \, w^2} + 16 \, \frac{w \sin(w \, t)}{64 \, w^4 + 64 - 127 \, w^2}$$

We can select the parts that contain $\cos(wt)$ and $\sin(wt)$ and compute the amplitude.

```
> select(has,steady,cos(w*t)):
```

```
> a:=simplify(")/cos(w*t);
```

$$a := -128 \, \frac{w^2 - 1}{64 \, w^4 + 64 - 127 \, w^2}$$

```
> select(has,steady,sin(w*t)):
```

```
> b:=simplify(")/sin(w*t);
```

$$b := 16 \, \frac{w}{64 \, w^4 + 64 - 127 \, w^2}$$

```
> a^2+b^2:
```

```
> ampsqr:=simplify(");
```

$$ampsqr := \frac{256}{64 \, w^4 + 64 - 127 \, w^2}$$

We maximize the amplitude by minimizing the denominator.

```
> T:=title='Amplitude Squared':
```

```
> plot(ampsqr,w=-2..2,T);
```

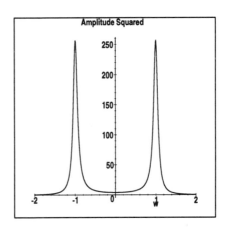

The maxima will be taken at the critical points.

```
> diff(ampsqr,w);
```

$$-256\,\frac{256\,w^3 - 254\,w}{(64\,w^4 + 64 - 127\,w^2)^2}$$

```
> sol:=solve(",{w});
```

$$sol := \{w = 0\},\ \{w = \frac{1}{16}\,\sqrt{254}\},\ \{w = -\frac{1}{16}\,\sqrt{254}\}$$

The maximum amplitude can then be computed.

```
> subs(sol[2],sqrt(ampsqr));
```

$$\frac{16}{255}\,\sqrt{256}\,\sqrt{255}$$

```
> evalf(");
```

$$16.03134185$$

4.3 Exercises

1. Investigate further what happens to the solution of $u'' + u = 2\cos(wt)$ as w approaches 1. Do the solutions to this equation in any sense approach the solution of $u'' + u = 2\cos t$?

2. Find the value of m that maximizes the amplitude of the steady-state solution of

$$mu'' + \frac{1}{8}u' + u = 2\cos(2t).$$

3. A 10 kg mass is attached to a spring having a spring constant of 140 N/m. The mass is started in motion from the equilibrium position with an initial velocity of 1 m/s in the upward direction and with an applied force $f(t) = 5 \sin t$. If the force due to air resistance is $-90u'$ N, find the subsequent motion of the mass. Plot its motion. Find the amplitude of the steady-state solution. If the applied force is $5 \sin(wt)$, find the value of w which maximizes the amplitude of the steady-state solution.

4. A 16-lb object is attached to a spring having a spring constant of 25 lb/ft. From rest, the object is pulled 7/6 ft below equilibrium and released. The system is surrounded by a medium offering a resistance in pounds equal to u' and a force $f(t) = 305 \cos(4t)$ is applied to the system. Find the subsequent motion of the object and plot it. Find the amplitude of the steady-state solution.

Chapter 5

Laplace Transforms

Consider a simple R-L series electrical circuit with a 9 volt battery switched as follows: at time $t = 0$ the battery is connected, with the initial current $i(0) = 0$. One second later the polarity of the battery is reversed, and at time $t = 2$ the battery is removed and the circuit is closed. The differential equation of this circuit is given by

$$L\frac{di}{dt} + Ri = E(t), \quad i(0) = 0,$$

where

$$E(t) = \begin{cases} 9 & \text{if } 0 \le t < 1, \\ -9 & \text{if } 1 \le t < 2, \\ 0 & \text{if } t \ge 2. \end{cases}$$

In Release 4, we can enter the applied voltage into Maple using the `piecewise` command.

```
> restart:
```

```
> E:=piecewise(t<0,0,0<=t and t<1,9,1<=t and t<2,-9,t>2,0);
```

$$E := \begin{cases} 0 & t < 0 \\ 9 & -t \le 0 \text{ and } t - 1 < 0 \\ -9 & 1 - t \le 0 \text{ and } t - 2 < 0 \\ 0 & 2 < t \end{cases}$$

Then we enter the differential equation, including the `piecewise` expression.

```
> diffeq1:=L*diff(i(t),t)+R*i(t)=E;
```

$$\textit{diffeq1} := L\left(\frac{\partial}{\partial t}\, i(t)\right) + R\,i(t) = \begin{cases} 0 & t < 0 \\ 9 & -t \le 0 \text{ and } t - 1 < 0 \\ -9 & 1 - t \le 0 \text{ and } t - 2 < 0 \\ 0 & 2 < t \end{cases}$$

```
> inits:=i(0)=0;
```

$$\textit{inits} := i(0) = 0$$

We can then use `dsolve` to find the solution.

```
> sol1:=dsolve({diffeq1,inits},i(t));
```

$$sol1 := i(t) = \begin{cases} 0 & t \le 0 \\ \dfrac{9}{R} - 9\dfrac{e^{(-\frac{Rt}{L})}}{R} & t \le 1 \\ 18\dfrac{e^{(-\frac{R(t-1)}{L})}}{R} - \dfrac{9}{R} - 9\dfrac{e^{(-\frac{Rt}{L})}}{R} & t \le 2 \\ 18\dfrac{e^{(-\frac{R(t-1)}{L})}}{R} - 9\dfrac{e^{(-\frac{R(t-2)}{L})}}{R} - 9\dfrac{e^{(-\frac{Rt}{L})}}{R} & 2 < t \end{cases}$$

The Laplace transform is a tool which simplifies this process. It also frequently enables `dsolve` to give nicer solutions.

5.1   Definition of the Laplace Transform

The *Laplace transform* of a function $f : [0, \infty) \to \mathcal{R}$ is defined by

$$F(s) = \mathcal{L}\{f(t)\}(s) := \int_0^\infty f(t)e^{-st}\,dt.$$

The integral is an improper integral, defined by

$$\lim_{A \to \infty} \int_0^A f(t)e^{-st}\,dt.$$

For example, compute the transform of $f(t) = 1$:

$$\begin{aligned} \mathcal{L}\{1\}(s) &= \lim_{A \to \infty} \int_0^A e^{-st}\,dt \\ &= \lim_{A \to \infty} \left. \frac{e^{-st}}{-s} \right|_0^A \\ &= \frac{1}{s} \quad \text{if } s > 0. \end{aligned}$$

Let's use Maple to compute the integral. We must tell Maple that $s > 0$ by using the `assume` command (otherwise the integral diverges):

```
> assume(s>0):
```

```
> Int(exp(-s*t),t=0..infinity);
```

$$\int_0^\infty e^{(-s^\sim t)}\,dt$$

```
> value(");
```

$$\frac{1}{s^\sim}$$

The tilde (˜)

indicates that an assumption has been placed on s. When you see a variable with a tilde, you can find out what assumptions have been made about the variable by using the about command.

The Laplace transform is accessed with the Maple package inttrans (integral transforms) via the command laplace. Before using Maple to compute the transform, remove the assumption that $s > 0$ (which also removes the tilde).

```
>  s:='s';
```
$$s := s$$

```
>  with(inttrans):
>  laplace(1,t,s);
```
$$\frac{1}{s}$$

Here are some additional examples:

```
>  laplace(exp(a*t),t,s);
```
$$\frac{1}{s - a}$$

```
>  laplace(sin(a*t),t,s);
```
$$\frac{a}{s^2 + a^2}$$

```
>  laplace(exp(a*t)*sin(b*t),t,s);
```
$$\frac{b}{(s - a)^2 + b^2}$$

This illustrates a general property:
$$\mathcal{L}\{e^{ct}f(t)\} = F(s - c).$$

5.2   Properties of the Laplace Transform

Although Laplace transforms are defined as improper integrals, they are not usually computed directly from the integral definition. Instead, one begins with a table of Laplace transforms of certain simple functions like 1, e^{at}, and $\sin t$, and one builds the Laplace transform of more complicated expressions from these simple examples by using linearity and certain rules. The manipulation rules are derived from the integral definition, usually by integration by parts or a simple substitution. For example, if an expression $\sin(2t)$ is multiplied by t^3, the Laplace transform is obtained by by multiplying the original Laplace transform by $(-1)^3$ and taking the third derivative with respect to s. Thus, we can compute the Laplace transform of $t^3 \sin 2t$ from a "table look up," (done here as the first Maple step), followed by a manipulation.

```
>  laplace(sin(2*t),t,s);
```
$$\frac{2}{s^2 + 4}$$

```
>   (-1)^3*diff(",s$3);
```

$$96\,\frac{s^3}{(s^2+4)^4} - 48\,\frac{s}{(s^2+4)^3}$$

```
>   simplify(");
```

$$48\,\frac{s\,(s^2-4)}{(s^2+4)^4}$$

It can also be done directly with Maple.

```
>   laplace(t^3*sin(2*t),t,s):
```

```
>   simplify(");
```

$$48\,\frac{s\,(s^2-4)}{(s^2+4)^4}$$

Similarly, if an expression $\sin 3t$ is multiplied by e^{at}, the Laplace transform is shifted to the right by a units, i.e., the variable s is replaced by $s-a$. For example, to take the Laplace transform of $t^3 e^{4t} \sin 5t$, we can begin with a "table look up" (the first step) and proceed as follows:

```
>   laplace(sin(5*t),t,s);
```

$$\frac{5}{s^2+25}$$

```
>   (-1)^3*diff(",s$3);
```

$$240\,\frac{s^3}{(s^2+25)^4} - 120\,\frac{s}{(s^2+25)^3}$$

```
>   simplify(");
```

$$120\,\frac{s\,(s^2-25)}{(s^2+25)^4}$$

```
>   subs(s=s-4,");
```

$$120\,\frac{(s-4)\,((s-4)^2-25)}{((s-4)^2+25)^4}$$

```
>   simplify(");
```

$$120\,\frac{(s-4)\,(s^2-8s-9)}{(s^2-8s+41)^4}$$

Alternatively, we could do the following:

```
>   laplace(sin(5*t),t,s);
```

$$\frac{5}{s^2+25}$$

```
>   subs(s=s-4,");
```

$$\frac{5}{(s-4)^2 + 25}$$

```
>   (-1)^3*diff(",s$3);
```

$$30 \frac{(2s-8)^3}{((s-4)^2+25)^4} - 60 \frac{2s-8}{((s-4)^2+25)^3}$$

```
>   simplify(");
```

$$120 \frac{(s-4)(s^2-8s-9)}{(s^2-8s+41)^4}$$

We can also go directly to Maple's Laplace transform.

```
>   laplace(t^3*exp(4*t)*sin(5*t),t,s);
```

$$240 \frac{(s-4)^3}{((s-4)^2+25)^4} - 120 \frac{s-4}{((s-4)^2+25)^3}$$

```
>   simplify(");
```

$$120 \frac{(s-4)(s^2-8s-9)}{(s^2-8s+41)^4}$$

Similarly, the process of going from a known Laplace transform back to the expression of which it is a transform, is usually done by table look up. In the process, a complicated fraction in the variable s is first converted to simple fractions in s by expressing the original in partial fraction form. The simple fractions which have linear denominators are looked up directly in the table. Those that have irreducible quadratic denominators are looked up after completing the square. For example, to find the inverse Laplace transform of

$$\frac{2s^2 - 6s + 8}{(s^2 - 2s + 5)(s - 3)},$$

we would begin by converting to partial fractions:

```
>   (2*s^2-6*s+8)/((s^2-2*s+5)*(s-3));
```

$$\frac{2s^2 - 6s + 8}{(s^2 - 2s + 5)(s - 3)}$$

```
>   convert(",parfrac,s);
```

$$\frac{1}{s-3} + \frac{-1+s}{s^2 - 2s + 5}$$

Then the first term would be a "table lookup;"

```
>   invlaplace(1/(s-3),s,t);
```

$$e^{(3t)}$$

and the second would proceed by completing the square:

```
>  with(student):
```

```
>  (s-1)/(s^2-2*s+5);
```

$$\frac{-1+s}{s^2 - 2s + 5}$$

```
>  completesquare(denom("));
```

$$(-1+s)^2 + 4$$

Once the denominator has been written as a power of $(s-1)$, and the numerator has been expressed that way, too, then a translation can be used, followed by a "table look up."

```
>  (s-1)/((s-1)^2+4);
```

$$\frac{-1+s}{(-1+s)^2 + 4}$$

```
>  subs(s-1=s,");
```

$$\frac{s}{s^2 + 4}$$

```
>  invlaplace(",s,t);
```

$$\cos(2\,t)$$

```
>  exp(t)*";
```

$$e^t \cos(2\,t)$$

The last step follows since shifting right by 1 in the s-domain corresponds to multiplication by e^t in the t-domain. Compare our results with Maple's inverse computation:

```
>  expr:=(2*s^2-6*s+8)/((s^2-2*s+5)*(s-3));
```

$$expr := \frac{2s^2 - 6s + 8}{(s^2 - 2s + 5)(s - 3)}$$

```
>  invlaplace(expr,s,t);
```

$$e^{(3t)} + e^t \cos(2\,t)$$

5.3 Derivatives

Laplace transforms are useful in solving initial value problems owing to the following property:

$$\mathcal{L}\{f^{(n)}(t)\} = s^n F(s) - s^{n-1} f(0) - \cdots - s f^{(n-2)}(0) - f^{(n-1)}(0),$$

i.e., *taking a derivative with respect to t is almost the same as multiplying the Laplace transform by s.* In other words, derivative operations are transformed into algebraic operations.

Let's examine this empirically using Maple. (Here `laplace(y(t),t,s)` is aliased to Y(s) for more pleasing output.)

```
>  alias(Y(s)=laplace(y(t),t,s)):

>  laplace(diff(y(t),t),t,s);
```
$$s\,Y(s) - y(0)$$

```
>  laplace(diff(y(t),t$2),t,s);
```
$$s\,(s\,Y(s) - y(0)) - D(y)(0)$$

```
>  laplace(diff(y(t),t$3),t,s);
```
$$s\,(s\,(s\,Y(s) - y(0)) - D(y)(0)) - (D^{(2)})(y)(0)$$

```
>  expand(");
```
$$s^3\,Y(s) - s^2\,y(0) - s\,D(y)(0) - (D^{(2)})(y)(0)$$

5.4 Initial Value Problems

To illustrate the use of the Laplace transform in solving initial value problems we present a few examples. Solve

$$y'' + y = \sin 2t, \quad y(0) = 2, \quad y'(0) = 1. \tag{5.1}$$

Here are the steps that we do:

1. Take the Laplace transform of the equation. (Since the transform is *linear*, this is just the sum of the Laplace transforms of each term.)

2. Solve the *algebraic* equation for the Laplace transform Y of the solution.

3. Convert the right hand side of the above result to partial fractions.

4. Recognize the individual terms in the partial fraction expansion as Laplace transforms of combinations of known functions.

5. Take the inverse Laplace transform of the equation, i.e., write down what $y(t)$ is, in terms of the known functions.

Observe that the differential equation (5.1) is transformed into a linear algebraic equation by the Laplace transform.

> `diffeq2:=diff(y(t),t$2)+y(t)=sin(2*t);`

$$diffeq2 := (\frac{\partial^2}{\partial t^2} \, y(t)) + y(t) = \sin(2\, t)$$

> `inits:=y(0)=2,D(y)(0)=1;`

$$inits := y(0) = 2, \, D(y)(0) = 1$$

> `laplace(diffeq2,t,s);`

$$s\, (s\, Y(s) - y(0)) - D(y)(0) + Y(s) = \frac{2}{s^2 + 4}$$

> `expand(");`

$$s^2\, Y(s) - s\, y(0) - D(y)(0) + Y(s) = \frac{2}{s^2 + 4}$$

> `subs(inits,");`

$$s^2\, Y(s) - 2\, s - 1 + Y(s) = \frac{2}{s^2 + 4}$$

> `Y(s)=solve(",Y(s));`

$$Y(s) = \frac{2\, s^3 + 8\, s + s^2 + 6}{s^4 + 5\, s^2 + 4}$$

> `convert(",parfrac,s);`

$$Y(s) = -\frac{2}{3} \frac{1}{s^2 + 4} + \frac{1}{3} \frac{5 + 6\, s}{s^2 + 1}$$

> `expand(");`

$$Y(s) = -\frac{2}{3} \frac{1}{s^2 + 4} + \frac{5}{3} \frac{1}{s^2 + 1} + 2 \frac{s}{s^2 + 1}$$

> `sol2a:=invlaplace(",s,t);`

$$sol2a := y(t) = -\frac{1}{3} \sin(2\, t) + \frac{5}{3} \sin(t) + 2 \cos(t)$$

The same operations can be done automatically in `dsolve` by inserting the `laplace` option.

> `sol2:=dsolve({diffeq2,inits},y(t),laplace);`

$$sol2 := y(t) = -\frac{1}{3} \sin(2\, t) + \frac{5}{3} \sin(t) + 2 \cos(t)$$

```
>  plot(rhs(sol2),t=0..2*Pi);
```

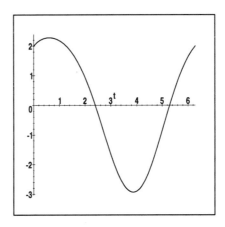

Here is another example, a fourth order equation. Solve

$$y^{(4)} - y = 0, \quad y(0) = 0, \quad y'(0) = 1, \quad y''(0) = 0, \quad y'''(0) = 0.$$

```
>  diffeq3:=diff(y(t),t$4)-y(t)=0;
```

$$diffeq3 := (\frac{\partial^4}{\partial t^4}\, y(t)) - y(t) = 0$$

```
>  inits:=y(0)=0,D(y)(0)=1,(D@@2)(y)(0)=0,(D@@3)(y)(0)=0;
```

$$inits := y(0) = 0,\, D(y)(0) = 1,\, (D^{(2)})(y)(0) = 0,\, (D^{(3)})(y)(0) = 0$$

```
>  laplace(diffeq3,t,s);
```

$$s\,(s\,(s\,(s\,Y(s) - y(0)) - D(y)(0)) - (D^{(2)})(y)(0)) - (D^{(3)})(y)(0) - Y(s) = 0$$

```
>  expand(");
```

$$s^4\,Y(s) - s^3\,y(0) - s^2\,D(y)(0) - s\,(D^{(2)})(y)(0) - (D^{(3)})(y)(0) - Y(s) = 0$$

```
>  subs(inits,");
```

$$s^4\,Y(s) - s^2 - Y(s) = 0$$

```
>  Y(s)=solve(",Y(s));
```

$$Y(s) = \frac{s^2}{s^4 - 1}$$

```
>  convert(",parfrac,s);
```

$$Y(s) = \frac{1}{4}\frac{1}{-1+s} - \frac{1}{4}\frac{1}{s+1} + \frac{1}{2}\frac{1}{s^2+1}$$

```
>  invlaplace(",s,t);
```

$$y(t) = \frac{1}{4}e^t - \frac{1}{4}e^{(-t)} + \frac{1}{2}\sin(t)$$

Alternatively,

```
>  sol3:=dsolve({diffeq3,inits},y(t),laplace);
```

$$sol3 := y(t) = \frac{1}{4}e^t - \frac{1}{4}e^{(-t)} + \frac{1}{2}\sin(t)$$

5.5   Unit Step Function

Let's return to the example at the start of this chapter. It described a simple R-L series electrical circuit with a switched 9 volt battery. The differential equation of this circuit is given by

$$L\frac{di}{dt} + Ri = E,$$

where

```
>  E:=piecewise(0<=t and t<1,9,1<=t and t<2,-9,t>=2,0);
```

$$E := \begin{cases} 9 & -t \le 0 \text{ and } t-1 < 0 \\ -9 & 1-t \le 0 \text{ and } t-2 < 0 \\ 0 & 2 \le t \end{cases}$$

While it would be easy to calculate the Laplace transform of E by direct integration, there is a simpler, more general method. Let $c \ge 0$ and define the unit step $u_c(t)$ by

$$u_c(t) = \begin{cases} 0 & \text{if } 0 \le t \le c, \\ 1 & \text{if } t > c. \end{cases}$$

With this definition, E is given as $E = 9 - 18u_1(t) + 9u_2(t)$. Maple can convert this for us:

```
>  convert(E,Heaviside);
```

$$9\,\text{Heaviside}(t) - 18\,\text{Heaviside}(t-1) + 9\,\text{Heaviside}(t-2)$$

In Maple, the unit step $u_0(t)$ is named Heaviside(t), so that $u_c(t) = $ Heaviside(t-c). (Note that replacing t by $t-c$ is a shift right by c, so that the unit jump is no longer at $t=0$.) The Laplace transform of $u_c(t)$ is

$$\int_0^\infty u_c(t)e^{-st}\,dt = \int_c^\infty e^{-st}\,dt = \frac{e^{-cs}}{s}.$$

Since the Laplace transform integral starts at $t=0$, the Laplace transform of the unit step is the same as the Laplace transform of the constant expression 1. The effect of shifting an expression right by c units in the t-domain is to multiply the Laplace transform by e^{cs}.

The unit step itself is only one of a class of expressions that arise naturally in electrical problems. Other useful expressions are easily represented using simple algebraic combinations of u_c. For example, the ramp, obtained by charging a capacitor at constant current, can be represented as follows:

```
>   f:=t*(1-Heaviside(t-1));
```

$$f := t\,(1 - \text{Heaviside}(t - 1))$$

```
>   plot(f,t=0..2,scaling=constrained,discont=true);
```

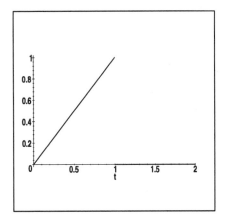

Here is an example of a differential equation involving a ramp.

```
>   diffeq4:=diff(y(t),t$2)+y(t)=f;
```

$$diffeq4 := (\frac{\partial^2}{\partial t^2}\,y(t)) + y(t) = t\,(1 - \text{Heaviside}(t - 1))$$

```
>   inits:=y(0)=0,D(y)(0)=0;
```

$$inits := y(0) = 0,\ D(y)(0) = 0$$

```
>   sol4:=dsolve({diffeq4,inits},y(t),laplace);
```

$$sol4 := y(t) =$$
$$t - \sin(t) + \text{Heaviside}(t - 1)\cos(t - 1) - t\,\text{Heaviside}(t - 1) + \text{Heaviside}(t - 1)\sin(t - 1)$$

Here is a plot of the solution.

```
>   plot(rhs(sol4),t=0..8);
```

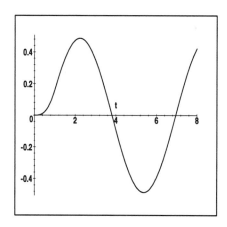

Another example: a constant voltage which is suddenly turned off after $t = 6.28$ seconds

>   `f:=1-Heaviside(t-2*Pi);`

$$f := 1 - \text{Heaviside}(t - 2\pi)$$

>   `plot(f,t=0..7,scaling=constrained);`

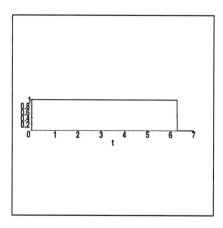

Let's solve the second order differential equation:

>   `diffeq5:=diff(y(t),t$2)+diff(y(t),t)+5/4*y(t)=1-Heaviside(t-Pi);`

$$\textit{diffeq5} := (\frac{\partial^2}{\partial t^2}\, y(t)) + (\frac{\partial}{\partial t}\, y(t)) + \frac{5}{4}\, y(t) = 1 - \text{Heaviside}(t - \pi)$$

```
>  inits:=y(0)=0,D(y)(0)=0;
```

$$inits := y(0) = 0, \; D(y)(0) = 0$$

```
>  sol5:=dsolve({diffeq5,inits},y(t),laplace);
```

$$sol5 := y(t) = \frac{4}{5} - \frac{4}{5} e^{(-1/2\,t)} \cos(t) - \frac{2}{5} e^{(-1/2\,t)} \sin(t) - \frac{4}{5} \mathrm{Heaviside}(t - \pi)$$
$$- \frac{4}{5} \mathrm{Heaviside}(t - \pi) \, e^{(-1/2\,t+1/2\,\pi)} \cos(t) - \frac{2}{5} \mathrm{Heaviside}(t - \pi) \, e^{(-1/2\,t+1/2\,\pi)} \sin(t)$$

Notice that the solution is complicated, which is to be expected with the discontinuous forcing function.

```
>  plot(rhs(sol5),t=0..6);
```

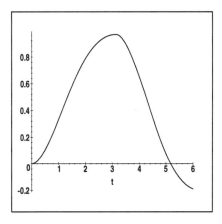

Now, let's take the derivative of this solution:

```
>  diff(rhs(sol5),t);
```

$$e^{(-1/2\,t)} \sin(t) - \frac{4}{5} \mathrm{Dirac}(t - \pi) - \frac{4}{5} \mathrm{Dirac}(t - \pi) \, e^{(-1/2\,t+1/2\,\pi)} \cos(t)$$
$$+ \mathrm{Heaviside}(t - \pi) \, e^{(-1/2\,t+1/2\,\pi)} \sin(t) - \frac{2}{5} \mathrm{Dirac}(t - \pi) \, e^{(-1/2\,t+1/2\,\pi)} \sin(t)$$

There is a new function in this formula. `Dirac` is the Maple name of the *impulse* $\delta(t)$. It arises here as the derivative of `Heaviside`. (See Exercise 6.)

Still another example: a truncated ramp

```
>  g:=t-Heaviside(t-Pi/2)*(t-Pi/2);
```

$$g := t - \mathrm{Heaviside}(t - \frac{1}{2}\pi)\,(t - \frac{1}{2}\pi)$$

```
> plot(g,t=0..3,scaling=constrained);
```

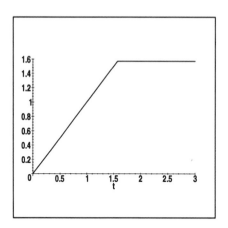

```
> diffeq6:=diff(y(t),t$2)+4*y(t)=g;
```

$$diffeq6 := (\frac{\partial^2}{\partial t^2}\, y(t)) + 4\,y(t) = t - \text{Heaviside}(t - \frac{1}{2}\,\pi)\,(t - \frac{1}{2}\,\pi)$$

```
> inits:=y(0)=0,D(y)(0)=0;
```

$$inits := y(0) = 0,\ \text{D}(y)(0) = 0$$

```
> sol6:=dsolve({diffeq6,inits},y(t),laplace);
```

$$sol6 := y(t) = \frac{1}{4}\,t - \frac{1}{8}\,\sin(2\,t) - \frac{1}{4}\,\text{Heaviside}(t - \frac{1}{2}\,\pi)\,t + \frac{1}{8}\,\text{Heaviside}(t - \frac{1}{2}\,\pi)\,\pi$$
$$- \frac{1}{8}\,\text{Heaviside}(t - \frac{1}{2}\,\pi)\,\sin(2\,t)$$

```
> plot(rhs(sol6),t=0..8);
```

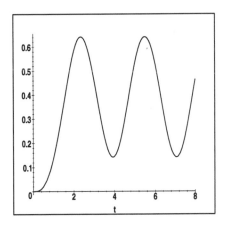

5.6 Impulse Functions

Consider the expression $I(h, t)$ defined by

$$I(h, t) = \begin{cases} 1/h & \text{if } 0 \le t \le h, \\ 0 & \text{otherwise.} \end{cases}$$

Another representation of I is $(1 - u_h(t))/h$. We can compute the Laplace transform of I, and obtain

$$\mathcal{L}\{I(h, t)\}(s) = \frac{1 - e^{-hs}}{hs} \to 1 \quad \text{as } h \to 0.$$

While the limit of I as $h \to 0$ does not exist at $t = 0$, it is still convenient to regard this limit as a "function," denoted by $\delta(t)$, with Maple name `Dirac`. A number of properties can be deduced:

- $\displaystyle\int_0^\infty \delta(t)\, dt = 1.$

- $\delta(t) = 0, \quad t \neq 0.$

- $\mathcal{L}\{\delta(t - t_0)\}(s) = e^{-st_0}, \quad t_0 \ge 0.$

- $\delta(t - t_0) = \dfrac{du_{t_0}}{dt}(t).$ (See Exercise 7.)

- $\displaystyle\int_0^\infty \delta(t - t_0) f(t)\, dt = f(t_0).$ (See Exercise 8.)

It is clear that no *function* (in the classical sense) can satisfy all these properties. However, for second order differential equations, there is a connection illustrated by the following examples.

```
>  diffeq7:=diff(y(t),t$2)+2*diff(y(t),t)+2*y(t)=Dirac(t);
```

$$diffeq7 := (\frac{\partial^2}{\partial t^2} y(t)) + 2\,(\frac{\partial}{\partial t} y(t)) + 2\,y(t) = \text{Dirac}(t)$$

```
> inits:=y(0)=0,D(y)(0)=0:
```

```
> sol7:=dsolve({diffeq7,inits},y(t),laplace);
```

$$sol7 := y(t) = e^{(-t)}\sin(t)$$

```
> diffeq8:=diff(y(t),t$2)+2*diff(y(t),t)+2*y(t)=0;
```

$$diffeq8 := (\frac{\partial^2}{\partial t^2}y(t)) + 2\,(\frac{\partial}{\partial t}y(t)) + 2\,y(t) = 0$$

```
> inits:=y(0)=0,D(y)(0)=1:
```

```
> sol8:=dsolve({diffeq8,inits},y(t),laplace);
```

$$sol8 := y(t) = e^{(-t)}\sin(t)$$

This example suggests that the second order equation $y'' + ay' + by = \delta(t)$, $y(0) = 0$, $y'(0) = 0$ is equivalent to $y'' + ay' + by = 0$, $y(0) = 0$, $y'(0) = 1$. If the equation were derived from modeling a mass-spring problem, then giving the mass an initial velocity is carried out physically by transferring momentum to the mass, e.g., by striking the mass with a hammer.

In the following example, we see `Dirac(t-Pi)`. This represents an impulse (the hammer blow), which occurs after $t = \pi$ seconds.

```
> laplace(Dirac(t-4),t,s);
```

$$e^{(-4\,s)}$$

```
> diffeq9:=diff(y(t),t$2)+2*diff(y(t),t)+2*y(t)=Dirac(t-Pi);
```

$$diffeq9 := (\frac{\partial^2}{\partial t^2}y(t)) + 2\,(\frac{\partial}{\partial t}y(t)) + 2\,y(t) = \mathrm{Dirac}(t - \pi)$$

```
> inits:=y(0)=0,D(y)(0)=0:
```

```
> sol9:=dsolve({diffeq9,inits},y(t),laplace);
```

$$sol9 := y(t) = -\mathrm{Heaviside}(t - \pi)\,e^{(-t+\pi)}\sin(t)$$

```
> plot(rhs(sol9),t=0..10);
```

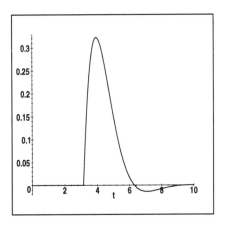

5.7   The Convolution Integral

The *convolution* of two functions f and g is defined to be the integral

$$(f * g)(t) = \int_0^t f(t - \tau) g(\tau) \, d\tau.$$

The *convolution theorem* states that

$$\mathcal{L}\{(f * g)(t)\}(s) = F(s)G(s).$$

That is, *the Laplace transform of the convolution is the product of the Laplace transforms.*
Maple knows this theorem:

```
>   unassign('f','g');
>   alias(F(s)=laplace(f(t),t,s)):
>   alias(G(s)=laplace(g(t),t,s)):
>   h:=int(f(t-tau)*g(tau),tau=0..t);
```

$$h := \int_0^t f(t - \tau)\, g(\tau)\, d\tau$$

```
>   laplace(h,t,s);
```

$$G(s)\, F(s)$$

The following integral is the convolution of t^2 and $\cos 2t$:

```
>   Int((t-tau)^2*cos(2*tau),tau=0..t);
```

$$\int_0^t (t - \tau)^2 \cos(2\,\tau)\, d\tau$$

```
>  f:=value(");
```

$$f := -\frac{1}{2}\sin(t)\cos(t) + \frac{1}{2}t$$

Take the Laplace transform:

```
>  laplace(f,t,s);
```

$$-\frac{1}{2}\frac{1}{s^2+4} + \frac{1}{2}\frac{1}{s^2}$$

```
>  simplify(");
```

$$\frac{2}{(s^2+4)s^2}$$

Now take the product of the individual transforms:

```
>  laplace(t^2,t,s)*laplace(cos(2*t),t,s);
```

$$\frac{2}{(s^2+4)s^2}$$

As you can see, the two transforms are the same.

5.8   The Transfer Function

The *transfer function* $H(s)$ of a linear system is defined as the ratio of the Laplace transforms of the "output" function $y(t)$ to the "input" function $f(t)$, assuming that all initial conditions are zero. In problems modeled by differential equations with constant coefficients, this function can be determined, as in the following example:

$$ay'' + by' + cy = f(t), \quad y(0) = y'(0) = 0,$$
$$(as^2 + bs + c)Y(s) = F(s),$$
$$H(s) = \frac{Y(s)}{F(s)} = \frac{1}{as^2 + bs + c}.$$

The *impulse response* is the inverse transform of the transfer function. It is well named, for if the input function $f(t)$ is $\delta(t)$, then the impulse response is simply the same as the inverse transform of $1/p(s)$, where p is the characteristic polynomial.

5.9   Exercises

1. Find the Laplace transform of $e^{-2t}\cos 3t$.

2. By using Euler's formula and complex algebra, we can obtain a number of transforms as special cases of a common formula. Show that

$$\mathcal{L}\{e^{(a+ib)t}\}(s) = \frac{s - a - ib}{(s-a)^2 + b^2}, \quad s > a.$$

3. By differentiating under the integral defining $F(s)$, show that

$$-\frac{dF}{ds}(s) = \mathcal{L}\{tf(t)\}(s)$$

$$\frac{d^2F}{ds^2}(s) = \mathcal{L}\{t^2 f(t)\}(s)$$

$$-\frac{d^3F}{ds^3}(s) = \mathcal{L}\{t^3 f(t)\}(s)$$

Generalize. Does Maple help?

4. Use the first formula in Exercise 3 to compute the transform of $t \cos t$.

5. Find the Laplace transform of

$$f(t) = \begin{cases} t & \text{if } 0 \le t < 1, \\ 2 - t & \text{if } 1 \le t < 2, \\ 0 & \text{if } t \ge 2. \end{cases}$$

6. Let $U_c(h, t)$ be defined by

$$U_c(h, t) = \begin{cases} 0 & \text{if } 0 \le t < c, \\ (t - c)/h & \text{if } c \le t < c + h, \\ 1 & \text{if } t \ge c + h. \end{cases}$$

Possibly using Maple, sketch the graph of $U_c(h, t)$.

(a) Show that

$$\lim_{h \to 0+} U_c(h, t) = u_c(t), \quad t \ne c.$$

(b) Show that

$$\frac{dU_c(h, t)}{dt} = I(h, t - c), \quad t \ne c.$$

(c) Discuss how this relates to the equation

$$\frac{du_c}{dt}(t) = \delta(t - c),$$

or, in Maple, `diff(Heaviside(t-c),t)=Dirac(t-c)`.

7. Use Maple to find the formal derivative of `Heaviside`$(t - t_0)$. Show that $\dfrac{du_{t_0}(t)}{dt} = 0$ if $t \ne t_0$.

8. Use the mean value theorem for integrals to show that

$$\lim_{h \to 0} \int_0^\infty I(h, t - t_0) f(t) \, dt = f(t_0)$$

at a point t_0 where f is continuous.

9. In Section 6.5, the solution to a nonhomogeneous first order initial value problem

$$\mathbf{x}'(t) = \mathbf{A}\mathbf{x}(t) + \mathbf{f}(t), \quad \mathbf{x}(0) = \mathbf{x0} \tag{5.2}$$

is given as

$$\mathbf{x}(t) = e^{\mathbf{A}t}\left[\mathbf{x0} + \int_0^t e^{-\mathbf{A}s}\mathbf{f}(s)\,ds\right].$$

(a) Show that the solution can be rewritten as

$$\mathbf{x}(t) = e^{\mathbf{A}t}\mathbf{x0} + \int_0^t e^{\mathbf{A}(t-s)}\mathbf{f}(s)\,ds.$$

Observe that the integral on the right hand side is a convolution.

(b) Derive the formula by taking the Laplace transform of the differential equation (5.2) and showing that

$$\mathbf{X}(s) = (s\mathbf{I} - \mathbf{A})^{-1}(\mathbf{F}(s) + \mathbf{x0}).$$

10. Your textbook lists a number of properties of `Dirac`. How many of these can you show in Maple?

11. The Laplace transform method is not restricted to single differential equations. Consider the initial value problem given by $y_1' = y_1 + 3y_2$, $y_2' = 3y_1 + y_2$ with initial conditions $y_1(0) = 3$, $y_2(0) = 1$. The Laplace transform changes the system of two simultaneous *differential* equations into a system of two simultaneous *algebraic* equations, which can be solved by traditional algebraic means.

```
>  with(inttrans):
>  alias(Y1(s)=laplace(y1(t),t,s),Y2(s)=laplace(y2(t),t,s)):
>  diffeq1:=diff(y1(t),t)=y1(t)+3*y2(t);
>  diffeq2:=diff(y2(t),t)=3*y1(t)+y2(t);
>  inits:=y1(0)=3,y2(0)=1;
>  laplace(diffeq1,t,s);eq1:=subs(inits,");
>  laplace(diffeq2,t,s);eq2:=subs(inits,");
>  ss:=solve({eq1,eq2},{Y1(s),Y2(s)});
>  rhs(ss[1]);convert(",parfrac,s);
>  sol1:=y1(t)=invlaplace(",s,t);
>  rhs(ss[2]);convert(",parfrac,s);
>  sol2:=y2(t)=invlaplace(",s,t);
```

Check that the solutions work:

```
>  subs({sol1,sol2},{diffeq1,diffeq2}):simplify(",exp);
>  subs(t=0,{sol1,sol2}):simplify(");
```

Chapter 6

Systems

The differential equations considered so far have involved only a single dependent variable. In this chapter we turn to problems involving a linear system of two or more differential equations with two or more dependent variables. It is sufficient to consider only systems of *first order* differential equations, since it is straightforward to convert higher order equations into this form. (See also Section 11.1.2.)

In order to handle more efficiently several linear equations in several unknowns, it will be appropriate to introduce some concepts from linear algebra, such as vectors, matrices, eigenvalues, and eigenvectors. This chapter will therefore include a section presenting Maple's linear algebra package.

The organization of this chapter is as follows:

- A few easy examples with dsolve, dsolve/numeric, and DEplot

- An example of how a system of two linear first order equations arises as a model of two tanks containing solutions of salt. We use Maple's dsolve command to solve this system, and DEplot to plot the direction field and several solutions.

- An introduction to Maple's linear algebra package.

- Using eigenvalues and eigenvectors to solve homogeneous first order linear systems.

- Computing matrix exponentials and solving nonhomogeneous first order linear systems.

6.1 dsolve, dsolve/numeric, and DEplot

Systems of simultaneous differential equations are more complicated than single equations, but often they may be solved directly or approximated numerically with the same Maple tools that we used for single equations, using essentially the same syntax.

Consider the system
$$x_1' = 2x_1 + x_2, \quad x_1(0) = 2,$$
$$x_2' = -4x_1 + 2x_2, \quad x_2(0) = 2.$$

```
>  diffeq11:=diff(x1(t),t)=2*x1(t)+x2(t);
```

$$diffeq11 := \frac{\partial}{\partial t} \text{x1}(t) = 2\,\text{x1}(t) + \text{x2}(t)$$

> `diffeq12:=diff(x2(t),t)=-4*x1(t)+2*x2(t);`

$$diffeq12 := \frac{\partial}{\partial t} \text{x2}(t) = -4\,\text{x1}(t) + 2\,\text{x2}(t)$$

To find the general solution to the ststem, we can try `dsolve`. (Note that the order of appearance of x1 and x2 in the output set is random.)

> `sys1g:=dsolve({diffeq11,diffeq12},{x1(t),x2(t)});`

$$sys1g := \{\text{x1}(t) = e^{(2\,t)} \cos(2\,t)\,\_C1 + \frac{1}{2} e^{(2\,t)} \sin(2\,t)\,\_C2,$$
$$\text{x2}(t) = -2\,e^{(2\,t)} \sin(2\,t)\,\_C1 + e^{(2\,t)} \cos(2\,t)\,\_C2\}$$

To solve an initial value problem, we include the initial data with the equations.

> `inits1:=x1(0)=2,x2(0)=2;`

$$inits1 := \text{x1}(0) = 2,\ \text{x2}(0) = 2$$

> `sys1p:=dsolve({diffeq11,diffeq12,inits1},{x1(t),x2(t)});`

$$sys1p := \{\text{x1}(t) = 2\,e^{(2\,t)} \cos(2\,t) + e^{(2\,t)} \sin(2\,t),\ \text{x2}(t) = -4\,e^{(2\,t)} \sin(2\,t) + 2\,e^{(2\,t)} \cos(2\,t)\}$$

To find floating point decimal values for the solutions at $t = 2$, we use:

> `subs(t=2,sys1p);`

$$\{\text{x2}(2) = -4\,e^4 \sin(4) + 2\,e^4 \cos(4),\ \text{x1}(2) = 2\,e^4 \cos(4) + e^4 \sin(4)\}$$

> `evalf(");`

$$\{\text{x2}(2) = 93.90459974,\ \text{x1}(2) = -112.6954811\}$$

To plot x_1 versus t over the interval $[0, 1.5]$, we use:

> `subs(sys1p,x1(t));`

$$2\,e^{(2\,t)} \cos(2\,t) + e^{(2\,t)} \sin(2\,t)$$

```
>   plot(",t=0..1.5);
```

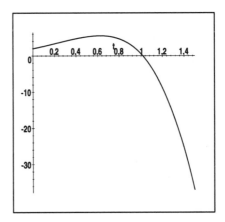

To make a parametric plot of x_1 versus x_2 for $-1 \le t \le 1$ we use:

```
>   plot([subs(sys1p,x1(t)),subs(sys1p,x2(t)),t=-1..1],view=[-2..2,-2..2]);
```

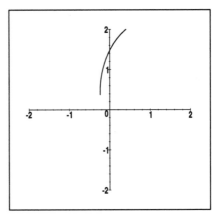

If dsolve is unable to solve the system, then we can approximate the solution, provided that we have *numerical initial values*. We use dsolve/numeric. Consider the system

$$x_1(t)' = \sin(x_1(t)x_2(t)), \quad x_1(0) = 0.5,$$
$$x_2(t)' = x_1(t) + x_2(t), \quad x_2(0) = 0.7.$$

```
>   diffeq21:=diff(x1(t),t)=sin(x1(t)*x2(t));
```

$$diffeq21 \; := \; \frac{\partial}{\partial t}\, \text{x1}(t) = \sin(\text{x1}(t)\,\text{x2}(t))$$

> `diffeq22:=diff(x2(t),t)=x1(t)+x2(t);`

$$diffeq22 \; := \; \frac{\partial}{\partial t}\, \text{x2}(t) = \text{x1}(t) + \text{x2}(t)$$

> `inits2:=x1(0)=0.5,x2(0)=0.7;`

$$inits2 \; := \; \text{x1}(0) = .5,\; \text{x2}(0) = .7$$

> `sys2:=dsolve({diffeq21,diffeq22,inits2},{x1(t),x2(t)},numeric);`

$$sys2 := \mathbf{proc}(rkf45\_x) \; \dots \; \mathbf{end}$$

To find the values of the solutions at $t = 0.3$, we use:

> `sys2(0.3);`

$$[t = .3,\; \text{x1}(t) = .6475521687485829,\; \text{x2}(t) = 1.141476861630875]$$

To plot x_1 versus t on the interval $[-1, 1.5]$, we use `plots[odeplot]`;

> `with(plots):`

> `odeplot(sys2,[t,x1(t)],-1..1.5);`

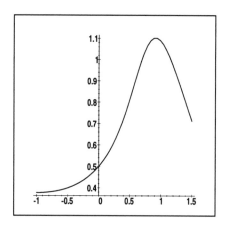

To plot a trajectory, i.e., a parametric plot of x_1 versus x_2 for t in $[-1, 1]$, we again use `odeplot`. Since it may be difficult to predict how far the curve will go for these values of t, we restrict the viewing area to the square $-2 \le x \le 2, -2 \le y \le 2$.

> `odeplot(sys2,[x1(t),x2(t)],-1..1,view=[-2..2,-2..2]);`

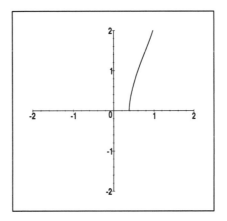

When the system is *autonomous*, i.e., the independent variable does not appear explicitly, we can get a direction field using DEplot.

```
>   with(DEtools):

>   DEplot({diffeq21,diffeq22},[x1(t),x2(t)],t=-1..1,x1=-2..2,x2=-2..2);
```

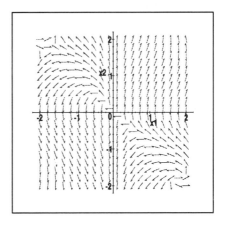

By including an initial value, we can get a trajectory on top of the arrows.

```
>   init:=[[x1(0)=0.5,x2(0)=0.7]]:

>   DEplot({diffeq21,diffeq22},[x1(t),x2(t)],t=-1..1,x1=-2..2,x2=-2..2,init);
```

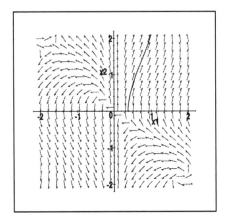

We can also plot the trajectory without the direction field. Note how we can shorten the length of the plot command by placing some of the arguments on previous lines, making the commands more readable. Also note that there is a short form for entering the initial values.

```
>   init:=[[0,0.5,0.7]]:
```

```
>   ranges:=t= -1..1,x1=-2..2,x2=-2..2:
```

```
>   DEplot({diffeq21,diffeq22},[x1(t),x2(t)],ranges,init);
```

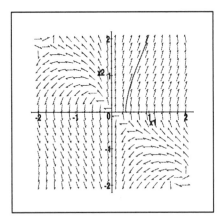

In the following sections we will study the classical solution methods, with Maple doing the arithmetic for us at intermediate steps, rather than having Maple run on autopilot. Situations can arise where we will be able to assist Maple to find a solution that it cannot find on its own, or where we will be able to guide Maple to find nicer solutions than it would find on its own.

6.2 Geometric Insight into Eigenvectors

Maple's `dsolve` command is useful for quickly finding solutions to *some* linear first order systems. However, as you will soon learn, solving first order linear systems is intimately related to finding roots of polynomials. Since there is no formula for the roots of polynomials of degree five or larger, you might guess that `dsolve` will be unable to solve many first order linear systems having five or more unknowns. At first glance it might appear that obtaining information about the solutions of large first order linear systems will be difficult. As it turns out, most of the needed information can be obtained using the notions of *eigenvalues* and *eigenvectors*.

What are these objects? Systems of first order differential equations offer an opportunity to see what they are and how they arise naturally in applications. We illustrate this with a model of two continuously stirred tanks containing salt solutions.

Consider two tanks, labeled 1 and 2, respectively. For $i = 1, 2$, tank i contains $Q_i(t)$ pounds of dissolved salt at time t, in a constant volume of V_i gallons. Hence, the concentration of salt in tank i is $c_i(t) = Q_i(t)/V_i$. Suppose that at time $t = 0$ fresh water begins flowing into tank 1 at the rate of r_1 gallons per minute, and mixed solution is pumped out at the same rate. In addition, solution from tank 1 is pumped into tank 2 at the rate of r_2 gallons per minute; and the mixed solution from tank 2 is returned to tank 1 at the same rate. Based on the principle that

time rate of change = (concentration in) × (flow rate in) − (concentration out) × (flow rate out)

we derive the following system of differential equations which describes the time rate of change of the amount of salt in tank i,

$$\frac{dQ_1}{dt} = -\frac{(r1+r2)Q1}{V1} + \frac{r2\,Q2}{V2}$$
$$\frac{dQ_2}{dt} = \frac{r2\,Q1}{V1} - \frac{r2\,Q2}{V2}$$

```
>  diffeq1:=diff(Q1(t),t)=-(r1+r2)*Q1(t)/V1+r2*Q2(t)/V2;
```

$$diffeq1 := \frac{\partial}{\partial t} Q1(t) = -\frac{(r1+r2)\,Q1(t)}{V1} + \frac{r2\,Q2(t)}{V2}$$

```
>  diffeq2:=diff(Q2(t),t)=r2*Q1(t)/V1-r2*Q2(t)/V2;
```

$$diffeq2 := \frac{\partial}{\partial t} Q2(t) = \frac{r2\,Q1(t)}{V1} - \frac{r2\,Q2(t)}{V2}$$

The overall effect of pumping fresh water in and mixed solution out is to wash the salt out of both tanks, so we expect the solutions $Q_i(t)$ to the above system of equations to have limiting value 0 as $t \to \infty$, for any starting values $Q_1(0)$ and $Q_2(0)$. We will use the parameter values $r_1 = 1, r_2 = 2, V_1 = 1, V_2 = 2/3$.

```
>  params:={r1=1,r2=2,V1=1,V2=2/3}:
```

```
>  diffeq1p:=subs(params,diffeq1);
```

$$diffeq1p := \frac{\partial}{\partial t} Q1(t) = -3\,Q1(t) + 3\,Q2(t)$$

```
>   diffeq2p:=subs(params,diffeq2);
```

$$diffeq2p := \frac{\partial}{\partial t}\,Q2(t) = 2\,Q1(t) - 3\,Q2(t)$$

We examine the direction field for the above system. We recall that DEplot is one way to do this.

```
>   with(DEtools):
```

```
>   DEplot([diffeq1p,diffeq2p],[Q1,Q2],t=0..1,Q1=0..1,Q2=0..1);
```

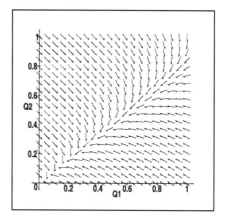

We observe from the plot that there seems to be a favored direction for solutions in the *Q1-Q2* plane (which is called the *phase plane*). We see next that, when we plot several solutions by giving several sets of initial conditions, the solutions do indeed appear to be attracted to one particular direction.

It turns out that this direction is that of an *eigenvector* for the 2×2 matrix $\begin{bmatrix} -3 & 3 \\ 2 & -3 \end{bmatrix}$ associated with the system of differential equations.

```
>   init1:=[0,0,1],[0,.5,1],[0,1,1]:
```

```
>   init2:=[0,1,.8],[0,1,.5],[0,1,0]:
```

```
>   inits:=init1,init2:
```

```
>   diffsp:=diffeq1p,diffeq2p:
```

```
>   Qrange:=Q1=0..1,Q2=0..1:
```

```
>   DEplot([diffsp],[Q1(t),Q2(t)],t=0..1,{inits},Qrange);
```

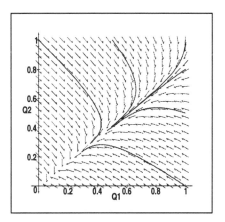

We next use Maple's `dsolve` command to find the general solution to the system.

```
> diffeq1p:=diff(Q1(t),t)=-3*Q1(t)+3*Q2(t);
```

$$diffeq1p := \frac{\partial}{\partial t} Q1(t) = -3\,Q1(t) + 3\,Q2(t)$$

```
> diffeq2p:=diff(Q2(t),t)=2*Q1(t)-3*Q2(t);
```

$$diffeq2p := \frac{\partial}{\partial t} Q2(t) = 2\,Q1(t) - 3\,Q2(t)$$

```
> interface(labelling=false):
```

```
> dsolve({diffeq1p,diffeq2p},{Q1(t),Q2(t)});
```

$$\{Q1(t) =$$
$$\frac{1}{2}\,\_C1\,e^{((-3+\sqrt{6})\,t)} + \frac{1}{2}\,\_C1\,e^{(-(3+\sqrt{6})\,t)} - \frac{1}{4}\,\_C2\,\sqrt{6}\,e^{(-(3+\sqrt{6})\,t)} + \frac{1}{4}\,\_C2\,\sqrt{6}\,e^{((-3+\sqrt{6})\,t)}$$
$$, Q2(t) = -\frac{1}{6}\,\_C1\,\sqrt{6}\,e^{(-(3+\sqrt{6})\,t)} + \frac{1}{6}\,\_C1\,\sqrt{6}\,e^{((-3+\sqrt{6})\,t)} + \frac{1}{2}\,\_C2\,e^{((-3+\sqrt{6})\,t)}$$
$$+ \frac{1}{2}\,\_C2\,e^{(-(3+\sqrt{6})\,t)}\}$$

The terms involving $e^{-(3+\sqrt{6})t}$ will die out faster than the other terms. A quick calculation shows that as t approaches infinity, the ratio $Q2(t)/Q1(t)$ approaches $\sqrt{2/3}$, or about 0.8165. This is the slope of the line representing the attracting direction in the above plots.

Now that we have a sense of what eigenvectors are, we can give a formal definition of an eigenvector of a system in terms of the eigenvectors of a matrix associated with that system. An eigenvector v for a matrix A is a nonzero vector that has the property that $Av = \lambda v$ for some scalar λ. When v is multiplied by the matrix A, it comes out a scalar multiple of itself. The number λ is called an *eigenvalue*.

This definition leads naturally into matrix arithmetic performed with the Maple `linalg` package described in the next section. Without going into detail about the commands, we use this package for one final computation: we will use Maple to find the eigenvectors and eigenvalues of the coefficient matrix in the above system.

```
>  with(linalg):
```

Warning, new definition for norm

Warning, new definition for trace

```
>  A:=matrix(2,2,[[-3,3],[2,-3]]);
```

$$A := \begin{bmatrix} -3 & 3 \\ 2 & -3 \end{bmatrix}$$

```
>  eigenvects(A);
```

$$[-3 + \sqrt{6}, \ 1, \ \{\left[1, \ \frac{1}{3}\sqrt{6}\right]\}], \ [-3 - \sqrt{6}, \ 1, \ \{\left[1, \ -\frac{1}{3}\sqrt{6}\right]\}]$$

The eigenvector $[1, \sqrt{6}/3]$, which has slope $\sqrt{6}/3 = \sqrt{2/3} \approx 0.8165$, is the one corresponding to the distinguished direction we have been observing. (In the output from `eigenvects`, the first entry in each list is an eigenvalue; the second entry is how often that eigenvalue occurs; and the third entry is an eigenvector.)

6.3   Maple's Linear Algebra Package

In Section 6.2 we introduced the notion of an eigenvector through its geometric appearance in the analysis of a mixing problem. We will need the Maple linear algebra package `linalg` to explore these notions further. We give a glimpse of this package below.

Start by loading the `linalg` package. (To suppress the list of commands in a package and keep a cleaner worksheet, use a colon.)

```
>  with(linalg);
```

[*BlockDiagonal, GramSchmidt, JordanBlock, LUdecomp, QRdecomp, Wronskian, addcol,*
  addrow, adj, adjoint, angle, augment, backsub, band, basis, bezout, blockmatrix, charmat,
  charpoly, cholesky, col, coldim, colspace, colspan, companion, concat, cond, copyinto,
  crossprod, curl, definite, delcols, delrows, det, diag, diverge, dotprod, eigenvals,
  eigenvalues, eigenvectors, eigenvects, entermatrix, equal, exponential, extend,
  ffgausselim, fibonacci, forwardsub, frobenius, gausselim, gaussjord, geneqns, genmatrix,
  grad, hadamard, hermite, hessian, hilbert, htranspose, ihermite, indexfunc, innerprod,
  intbasis, inverse, ismith, issimilar, iszero, jacobian, jordan, kernel, laplacian, leastsqrs,
  linsolve, matadd, matrix, minor, minpoly, mulcol, mulrow, multiply, norm, normalize,
  nullspace, orthog, permanent, pivot, potential, randmatrix, randvector, rank, ratform,
  row, rowdim, rowspace, rowspan, rref, scalarmul, singularvals, smith, stack, submatrix,
  subvector, sumbasis, swapcol, swaprow, sylvester, toeplitz, trace, transpose,
  vandermonde, vecpotent, vectdim, vector, wronskian]

If you have any experience with matrices and vectors, then there should be some familiar terms listed above. If you want specific information on any of these commands, simply type ?commandname at the prompt or use the Help menu.

Let's illustrate some of these commands. We will start with simple *vector/matrix* input, and progress to more complicated topics. We start by creating two vectors and manipulating them.

```
> uu:=vector([3,-1,0]);
```

$$uu := [3, -1, 0]$$

```
> vv:=vector([5,-2,14]);
```

$$vv := [5, -2, 14]$$

Notice what happens when we prompt Maple to echo either of these vectors:

```
> uu;vv;
```

$$uu$$

$$vv$$

Apparently nothing! To see **u** and **v** in their full glory, type the following.

```
> evalm(uu);evalm(vv);
```

$$[3, -1, 0]$$

$$[5, -2, 14]$$

The evalm command simply asks Maple to *evaluate the argument in its vector or matrix form*. If you are wondering whether you have to use this every time you work with matrices or vectors, the answer is "no." This command is only used to show the individual entries in a vector or matrix assigned to a variable. For example, if you want to add **u** and **v** and name the result **unew**, but you do not care to see the entries yet, then you can input

```
> unew:=uu+vv;
```

$$unew := uu + vv$$

If you want to see the result of this action, then simply type

```
> evalm(unew);
```

$$[8, -3, 14]$$

Should you feel the need to see everything as you create it, then use can use the command matadd. (This command is new to Release 4.)

```
> unew:=matadd(uu,vv);
```

$$unew := [8, -3, 14]$$

Subtracting vectors is just as simple.

```
> evalm(uu-vv);
```

$$[-2, 1, -14]$$

Alternatively, we can use the Release 4 `matadd` command with coefficients:

```
>  matadd(uu,vv,1,-1);
```

$$[-2, 1, -14]$$

Accessing particular entries in a vector is also easy.

```
>  uu[2];unew[3];
```

$$-1$$

$$14$$

Variable entries can also be placed into vectors.

```
>  w:=vector([x+y,y-2*x,sin(x),2*x-1,3,x*y*z]);
```

$$w := [x + y, y - 2x, \sin(x), 2x - 1, 3, xyz]$$

```
>  w[4];
```

$$2x - 1$$

Of course, the entries in vectors can be manipulated by functions, in ways similar to those in which scalar variables are manipulated, by *mapping* various commands onto the vector. For example, we can apply the exponential function to each component of w.

```
>  map(exp,w);
```

$$\left[e^{(x+y)}, e^{(y-2x)}, e^{\sin(x)}, e^{(2x-1)}, e^3, e^{(xyz)}\right]$$

(The exponential function has the property that it automatically maps onto vectors, as the following example shows. Not all functions have that property.)

```
>  exp(w);evalm(");
```

$$e^w$$

$$\left[e^{(x+y)}, e^{(y-2x)}, e^{\sin(x)}, e^{(2x-1)}, e^3, e^{(xyz)}\right]$$

Any additional parameters for a command are specified after the vector in the `map` command. For example, `diff` requires the parameter `,x`. Accordingly, the command

```
>  map(diff,w,x);
```

$$[1, -2, \cos(x), 2, 0, yz]$$

takes the partial derivative of each entry, and hence returns the partial derivative of the vector **w** with respect to x. Integration of a vector works the same way. (Of course, Maple still adds no constants of integration.)

```
>  map(int,",x);
```

$$[x, -2x, \sin(x), 2x, 0, xyz]$$

With map, the parameters follow the operand. Sometimes, the syntax of a command is such that the parameters precede the operand. New to Release 4 is the map2 command which handles this case. In the following command, subs is the operator, $x = 2$ is the parameter, and w is the operand. Since the parameter $x = 2$ precedes the operand w in subs (parameter, operand), we use map2.

```
> map2(subs,x=2,w);
```

$$[2 + y,\ y - 4,\ \sin(2),\ 3,\ 3,\ 2\,y\,z]$$

(Another way to achieve the same result is to substitute into the *evaluated form* of a vector.)

```
> subs(x=2,evalm(w));
```

$$[2 + y,\ y - 4,\ \sin(2),\ 3,\ 3,\ 2\,y\,z]$$

Note, however, that the subs command *does not* itself work on vectors.

```
> subs(x=2,w):evalm(");
```

$$[x + y,\ y - 2\,x,\ \sin(x),\ 2\,x - 1,\ 3,\ x\,y\,z]$$

Let's move on now to matrices. We will often use capital letters to refer to matrices, but this convention is definitely not mandatory. Let's create a few and use them below.

```
> A:=matrix(3,4,[2,-3,5,0,1,2,-1,3,0,2,-2,1]);
```

$$A := \begin{bmatrix} 2 & -3 & 5 & 0 \\ 1 & 2 & -1 & 3 \\ 0 & 2 & -2 & 1 \end{bmatrix}$$

```
> B:=matrix(3,3,[2,3,-3,1,1,-2,1,0,5]);
```

$$B := \begin{bmatrix} 2 & 3 & -3 \\ 1 & 1 & -2 \\ 1 & 0 & 5 \end{bmatrix}$$

```
> C:=matrix(3,3,[3,0,1,0,2,2,1,3,-2]);
```

$$C := \begin{bmatrix} 3 & 0 & 1 \\ 0 & 2 & 2 \\ 1 & 3 & -2 \end{bmatrix}$$

Adding or subtracting matrices is just like adding or subtracting vectors. To add **B** and **C**, either use

```
> matadd(B,C);
```

$$\begin{bmatrix} 5 & 3 & -2 \\ 1 & 3 & 0 \\ 2 & 3 & 3 \end{bmatrix}$$

or

```
> B+C:evalm(");
```

$$\begin{bmatrix} 5 & 3 & -2 \\ 1 & 3 & 0 \\ 2 & 3 & 3 \end{bmatrix}$$

It's also nice to know that Maple recognizes that matrices have to be the same size to be added or subtracted. For example,

```
> matadd(A,B);
```

`Error, (in matadd) matrix dimensions incompatible`

Maple also knows how to multiply matrices. **But beware!** Matrix multiplication is typically not commutative. That is, **BC** and **CB** are probably not the same.

```
> BC:=multiply(B,C);
```

$$BC := \begin{bmatrix} 3 & -3 & 14 \\ 1 & -4 & 7 \\ 8 & 15 & -9 \end{bmatrix}$$

```
> CB:=multiply(C,B);
```

$$CB := \begin{bmatrix} 7 & 9 & -4 \\ 4 & 2 & 6 \\ 3 & 6 & -19 \end{bmatrix}$$

In fact, Maple illustrates that they are not the same. Maple reserves the use of * for commutative multiplication and will give an error message if you try to multiply matrices with it. (The error message will only appear *after* you attempt to evaluate the result as a matrix!) Noncommutative multiplication is done using the &* operator.

```
> B&*C:evalm(");
```

$$\begin{bmatrix} 3 & -3 & 14 \\ 1 & -4 & 7 \\ 8 & 15 & -9 \end{bmatrix}$$

```
> C&*B:evalm(");
```

$$\begin{bmatrix} 7 & 9 & -4 \\ 4 & 2 & 6 \\ 3 & 6 & -19 \end{bmatrix}$$

These agree with the results of the `multiply` commands above.

Note that syntax such as `evalm(C&*")`; will not work, since Maple interprets the `&*"` as a new operator. You must use a space between the `&*` and the `"`. (Do a "Topic Search" in the "Help" facility on `&` to read about "neutral operator symbols.")

To compute the matrix product **(B)(C)(BC)(A)**, we can type

```
> multiply(B,C,BC,A);
```

$$
\begin{bmatrix}
449 & -138 & 587 & 534 \\
228 & -83 & 311 & 277 \\
-285 & 257 & -542 & -359
\end{bmatrix}
$$

and of course several matrix computations can be performed at once:

```
> evalm(B-B&*C+2*CB);
```

$$
\begin{bmatrix}
13 & 24 & -25 \\
8 & 9 & 3 \\
-1 & -3 & -24
\end{bmatrix}
$$

Maple's capabilities extend far beyond these. For example, determinants can be computed:

```
> det(B);
```

$$-8$$

Inverse matrices can be created and verified:

```
> 1/C:evalm(");
```

$$
\begin{bmatrix}
\dfrac{5}{16} & \dfrac{-3}{32} & \dfrac{1}{16} \\[2mm]
\dfrac{-1}{16} & \dfrac{7}{32} & \dfrac{3}{16} \\[2mm]
\dfrac{1}{16} & \dfrac{9}{32} & \dfrac{-3}{16}
\end{bmatrix}
$$

```
> invC:=inverse(C);
```

$$
invC := \begin{bmatrix}
\dfrac{5}{16} & \dfrac{-3}{32} & \dfrac{1}{16} \\[2mm]
\dfrac{-1}{16} & \dfrac{7}{32} & \dfrac{3}{16} \\[2mm]
\dfrac{1}{16} & \dfrac{9}{32} & \dfrac{-3}{16}
\end{bmatrix}
$$

```
>  multiply(C,invC);
```

$$\begin{bmatrix} 1 & 0 & 0 \\ 0 & 1 & 0 \\ 0 & 0 & 1 \end{bmatrix}$$

The student may happen across the strange notation &*(), which represents a "one size fits all" identity matrix. Observe its properties in the following results:

```
>  C&*(1/C);
```

$$C \,\&*\, \left(\frac{1}{C} \right)$$

```
>  evalm(");
```

$$\&*()$$

```
>  C+";evalm(");
```

$$C + \&*()$$

$$\begin{bmatrix} 4 & 0 & 1 \\ 0 & 3 & 2 \\ 1 & 3 & -1 \end{bmatrix}$$

```
>  diag(1$4)+&*();
```

$$\begin{bmatrix} 1 & 0 & 0 & 0 \\ 0 & 1 & 0 & 0 \\ 0 & 0 & 1 & 0 \\ 0 & 0 & 0 & 1 \end{bmatrix} + \&*()$$

```
>  evalm(");
```

$$\begin{bmatrix} 2 & 0 & 0 & 0 \\ 0 & 2 & 0 & 0 \\ 0 & 0 & 2 & 0 \\ 0 & 0 & 0 & 2 \end{bmatrix}$$

The fact that inverses can be computed indicates that Maple can be used to solve systems of equations. We demonstrate this in a number of ways below. Suppose we have a system of equations that we wish to solve.

```
> list_equations:=[3*x-y+z=-2,x-3*y+2*z=6,-x+y-3*z=5]:
```

One way to do this is simply to use Maple's `solve` command. First we need to convert this *list* to a *set*.

```
> set_equations:=convert(list_equations,set);
```

$$set\_equations := \{x - 3\,y + 2\,z = 6,\ -x + y - 3\,z = 5,\ 3\,x - y + z = -2\}$$

Notice that Maple did not return the equations in the same order that we typed them in. This is because Maple does not pay attention to the order of things in a `set`. We can now use the `solve` command.

```
> sol:=solve(set_equations,{x,y,z});
```

$$sol := \{z = \frac{-8}{3},\ x = \frac{-7}{6},\ y = \frac{-25}{6}\}$$

Since the answer appeared as a set, there is no natural order in which the variables will appear. Indeed, if the set of equations is solved a second time, the order in which the variables appear may not be the same as it was before. One way to force the numbers to come out in certain preferred order is to use substitution:

```
> subs(sol,[x,y,z]);
```

$$[\frac{-7}{6},\ \frac{-25}{6},\ \frac{-8}{3}]$$

We can also solve systems of equations by using matrix/vector methods. In this case, you can try typing in the augmented matrix for the system above; or you can use some of Maple's built in features. We will take the latter route. The following command forms a matrix from a list of linear equations, where the entries in the matrix are the respective coefficients of x, y, and z. The inclusion of the fourth argument, `flag`, forces Maple to append an extra column formed from the right hand sides of the equations in the list, i.e., it form the "augmented coefficient matrix."

```
> aug:=genmatrix(list_equations,[x,y,z],flag);
```

$$aug := \begin{bmatrix} 3 & -1 & 1 & -2 \\ 1 & -3 & 2 & 6 \\ -1 & 1 & -3 & 5 \end{bmatrix}$$

In Release 4, there is a command to reverse this operation. We can disassemble the matrix aug, using the submatrix command. The column matrix can be converted to a vector. The resulting coefficient matrix **A** and vector **b** are then used to form the linear equations.

```
> submatrix(aug,1..3,4..4):b:=convert(",vector);
```

$$b := [-2,\ 6,\ 5]$$

```
> A:=submatrix(aug,1..3,1..3);
```

$$A := \begin{bmatrix} 3 & -1 & 1 \\ 1 & -3 & 2 \\ -1 & 1 & -3 \end{bmatrix}$$

```
>  sys:=geneqns(A,[x,y,z],b);
```

$$sys := \{x - 3y + 2z = 6, -x + y - 3z = 5, 3x - y + z = -2\}$$

We can solve the system using the augmented matrix and some of the tools in the `linalg` package. For example, Maple has a command `rref` which produces the reduced row echelon form of a matrix.

```
>  rref(aug);
```

$$\begin{bmatrix} 1 & 0 & 0 & \dfrac{-7}{6} \\ 0 & 1 & 0 & \dfrac{-25}{6} \\ 0 & 0 & 1 & \dfrac{-8}{3} \end{bmatrix}$$

The solution generated by the `solve` command, agrees with the last column. Another way to solve this system is to rewrite the system in the form

$$\begin{bmatrix} 3 & -1 & 1 \\ 1 & -3 & 2 \\ -1 & 1 & -3 \end{bmatrix} \begin{bmatrix} x \\ y \\ z \end{bmatrix} = \begin{bmatrix} -2 \\ 6 \\ 5 \end{bmatrix}$$

In this case we have

```
>  coef_matrix:=delcols(aug,4..4);
```

$$coef\_matrix := \begin{bmatrix} 3 & -1 & 1 \\ 1 & -3 & 2 \\ -1 & 1 & -3 \end{bmatrix}$$

```
>  rhs_vector:=col(aug,4);
```

$$rhs\_vector := [-2, 6, 5]$$

The `linsolve` command can now be used to solve the system.

```
>  linsolve(coef_matrix,rhs_vector);
```

$$\left[\frac{-7}{6}, \frac{-25}{6}, \frac{-8}{3} \right]$$

If you are solving a large system, then using the `linsolve` command will produce the quickest results. It also has the nice property that it will find solutions of systems with multiple answers, for example, one equation in two unknowns or a system in which the second equation is a multiple of the first.

Let's next examine a few advanced commands. Consider the following matrix:

```
>  M:=matrix(3,3,[-49,33,-29,-25,69,-35,14,6,16]);
```

$$M := \begin{bmatrix} -49 & 33 & -29 \\ -25 & 69 & -35 \\ 14 & 6 & 16 \end{bmatrix}$$

The eigenvalues of **M** can be easily computed using the `eigenvals` command.

```
>  ev:=eigenvals(M);
```

$$ev := 18, -36, 54$$

Actually, it is possible to compute both eigenvalues and eigenvectors at the same time.

```
>  ev:=eigenvects(M);
```

$$ev := [-36, 1, \{[7, 1, -2]\}], [54, 1, \{[1, 4, 1]\}], [18, 1, \{[1, -5, -8]\}]$$

Note that the eigenvalues returned by `eigenvects` are the same numbers as those returned by `eigenvals`, but the order in which they appear need not be the same (as a matter of fact, the order in which they appear can change if you re-execute the command). Similarly, eigenvectors returned by `eigenvects` need not agree with those that you compute by hand, or by another method, or even by re-executing the command. Any multiple of an eigenvector is again an eigenvector; and different commands may return any scalar multiple of the eigenvector that you have.

If you have had any experience with matrices and eigenvalues, then you know that things rarely work out this simply. Usually eigenvalues and eigenvectors do not contain integers, even if the matrix has all integer entries. For example, consider

```
>  lseq1:=3,3,-2,1,0,3,2,3,1,1,-2,3,2:
>  lseq2:=1,-1,1,1,1,-3,1,0,1,0,1,-3:
>  lseq:=lseq1,lseq2:
>  N:=matrix(5,5,[lseq]);
```

$$N := \begin{bmatrix} 3 & 3 & -2 & 1 & 0 \\ 3 & 2 & 3 & 1 & 1 \\ -2 & 3 & 2 & 1 & -1 \\ 1 & 1 & 1 & -3 & 1 \\ 0 & 1 & 0 & 1 & -3 \end{bmatrix}$$

```
>  ev:=eigenvals(N);
```

$$ev := \mathrm{RootOf}(698 + 470\,\_Z - 14\,\_Z^2 - 44\,\_Z^3 - \_Z^4 + \_Z^5)$$

In this case, Maple is kind enough to tell us that the eigenvalues are the roots of a polynomial (the characteristic polynomial). Yet, unlike the preceding example, it failed to return the individual eigenvalues themselves. That's because the eigenvalues are not nice numbers like 3 or −2 like the entries in the matrix.

Maple can still find the eigenvalues, it just needs to approximate. One way to tell Maple that it is acceptable to approximate is to give it a matrix with floating point decimal entries instead of integers or ratios of integers. One way to do this is to is the `map` command. For example, `map(evalf,N)` converts all entries in a matrix to floats, so that the `eigenvals` and `eigenvects` commands will work. A second way is shown below.

```
>  ev:=evalf(Eigenvals(N));
```

$$ev := [6.008505572, 4.431627930, -4.633631957, -2.059186113, -2.747315442]$$

It is also possible to obtain approximate eigenvectors using a modification of this command.

```
>  eigmat:='eigmat':evalf(Eigenvals(N,eigmat));
```

$$[6.008505572, 4.431627930, -4.633631957, -2.059186113, -2.747315442]$$

```
>  evalm(eigmat);
```

$$\begin{bmatrix} -.4900793187\,, .5638315171\,, .5040840759\,, .4040678905\,, .3514793504 \\ -.5785126616\,, -.1683591841\,, -.5222483988\,, -.3606960366\,, -.6063687568 \\ -.2057891781\,, -.6769785455\,, .6667012181\,, .3677020734\,, .4041913723 \\ -.1504467269\,, -.04168258174\,, -.9478446825\,, -.2267624041\,, .6074263134 \\ -.08091901436\,, -.02826322441\,, .8998924645\,, -.6244151436\,, .0041852821 \end{bmatrix}$$

The columns of the preceding matrix are approximate eigenvectors corresponding to the eigenvalues previously listed. The ordering of the eigenvectors is the same as that of the eigenvalues. That is, the first column of the matrix *eigmat* is an approximate eigenvector corresponding to the approximate eigenvalue 6.00850557, etc. (**Caveat:** The reader who wants to use this command is referred to the Appendix for a discussion of its pecularities.)

Finally, let's see how to create matrices that have a distinct pattern by defining a procedure that operates on the row and column indices of the matrix. For example, suppose we need a 12×12 matrix whose diagonal entries are -2, super- and sub-diagonal entries are 1, and all other entries are 0. We do this as follows. First write a procedure which describes the entries in the matrix you want to build.

```
>  f:=proc(i,j)
   if i=j then -2
   elif i=j+1 then 1
   elif j=i+1 then 1
   else 0 fi
   end:
```

Then create the desired matrix by combining the `matrix` command and the procedure f.

```
>  P:=matrix(12,12,f);
```

$$
P := \begin{bmatrix}
-2 & 1 & 0 & 0 & 0 & 0 & 0 & 0 & 0 & 0 & 0 & 0 \\
1 & -2 & 1 & 0 & 0 & 0 & 0 & 0 & 0 & 0 & 0 & 0 \\
0 & 1 & -2 & 1 & 0 & 0 & 0 & 0 & 0 & 0 & 0 & 0 \\
0 & 0 & 1 & -2 & 1 & 0 & 0 & 0 & 0 & 0 & 0 & 0 \\
0 & 0 & 0 & 1 & -2 & 1 & 0 & 0 & 0 & 0 & 0 & 0 \\
0 & 0 & 0 & 0 & 1 & -2 & 1 & 0 & 0 & 0 & 0 & 0 \\
0 & 0 & 0 & 0 & 0 & 1 & -2 & 1 & 0 & 0 & 0 & 0 \\
0 & 0 & 0 & 0 & 0 & 0 & 1 & -2 & 1 & 0 & 0 & 0 \\
0 & 0 & 0 & 0 & 0 & 0 & 0 & 1 & -2 & 1 & 0 & 0 \\
0 & 0 & 0 & 0 & 0 & 0 & 0 & 0 & 1 & -2 & 1 & 0 \\
0 & 0 & 0 & 0 & 0 & 0 & 0 & 0 & 0 & 1 & -2 & 1 \\
0 & 0 & 0 & 0 & 0 & 0 & 0 & 0 & 0 & 0 & 1 & -2
\end{bmatrix}
$$

That sure beats typing in 144 entries!

(A second method of forming the matrix above is to use the band command in the linalg package.)

```
>  band([1,-2,1],12);
```

$$
\begin{bmatrix}
-2 & 1 & 0 & 0 & 0 & 0 & 0 & 0 & 0 & 0 & 0 & 0 \\
1 & -2 & 1 & 0 & 0 & 0 & 0 & 0 & 0 & 0 & 0 & 0 \\
0 & 1 & -2 & 1 & 0 & 0 & 0 & 0 & 0 & 0 & 0 & 0 \\
0 & 0 & 1 & -2 & 1 & 0 & 0 & 0 & 0 & 0 & 0 & 0 \\
0 & 0 & 0 & 1 & -2 & 1 & 0 & 0 & 0 & 0 & 0 & 0 \\
0 & 0 & 0 & 0 & 1 & -2 & 1 & 0 & 0 & 0 & 0 & 0 \\
0 & 0 & 0 & 0 & 0 & 1 & -2 & 1 & 0 & 0 & 0 & 0 \\
0 & 0 & 0 & 0 & 0 & 0 & 1 & -2 & 1 & 0 & 0 & 0 \\
0 & 0 & 0 & 0 & 0 & 0 & 0 & 1 & -2 & 1 & 0 & 0 \\
0 & 0 & 0 & 0 & 0 & 0 & 0 & 0 & 1 & -2 & 1 & 0 \\
0 & 0 & 0 & 0 & 0 & 0 & 0 & 0 & 0 & 1 & -2 & 1 \\
0 & 0 & 0 & 0 & 0 & 0 & 0 & 0 & 0 & 0 & 1 & -2
\end{bmatrix}
$$

6.4   Homogeneous First Order Linear Systems

This section contains a few examples which demonstrate how to find the solutions of first order linear initial value problems of the form

$$
\begin{aligned}
\mathbf{x}'(t) &= \mathbf{Ax}(t), & (6.1) \\
\mathbf{x}(0) &= \mathbf{x0}. & (6.2)
\end{aligned}
$$

We have chosen our examples to illustrate the solution process for three typical situations. In the first example, the coefficient matrix has distinct real eigenvalues. (Such an eventuality *guarantees* that there are sufficiently many eigenvectors. There are, however, examples like the identity matrix, in which eigenvalues are repeated, but there are still enough eigenvectors; and these fall into the first case, as well.) The coefficient matrix in the second example has *both* a repeated eigenvalue *and* a shortage of eigenvectors. Finally, the third coefficient matrix has real entries but complex eigenvalues.

6.4.1   Distinct eigenvalues

Consider the problem of trying to solve the initial value problem (6.1), (6.2), where $\mathbf{A} = \mathbf{A1}$ and $\mathbf{x0}$ are given by:

```
> with(linalg):

> A1:=matrix(3,3,[4,2,3,2,1,2,-1,2,0]);
```

$$
A1 := \begin{bmatrix} 4 & 2 & 3 \\ 2 & 1 & 2 \\ -1 & 2 & 0 \end{bmatrix}
$$

```
> x0:=matrix(3,1,[2,1,-1]);
```

$$
x0 := \begin{bmatrix} 2 \\ 1 \\ -1 \end{bmatrix}
$$

To solve this problem we need the eigenvalues and eigenvectors of $\mathbf{A1}$.

```
> ev:=eigenvects(A1);
```

$$
ev := [-1, 1, \{ \left[\frac{1}{2}, 1, \frac{-3}{2} \right] \}], [1, 1, \{[1, 0, -1]\}], [5, 1, \{[2, 1, 0]\}]
$$

Each eigenvector gives us a solution. (Note that we may choose *any* multiple of the eigenvector listed above, since a scalar multiple of an eigenvector is again an eigenvector. This freedom of choice will be compensated for in the choice of coefficient in the linear combination below.)

```
>  sol1:=exp(5*t)*matrix(3,1,[2,1,0]);
```

$$sol1 := e^{(5\,t)} \begin{bmatrix} 2 \\ 1 \\ 0 \end{bmatrix}$$

```
>  sol2:=exp(t)*matrix(3,1,[-1,0,1]);
```

$$sol2 := e^{t} \begin{bmatrix} -1 \\ 0 \\ 1 \end{bmatrix}$$

```
>  sol3:=exp(-t)*matrix(3,1,[1,2,-3]);
```

$$sol3 := e^{(-t)} \begin{bmatrix} 1 \\ 2 \\ -3 \end{bmatrix}$$

So the general solution to (6.1) is given by taking a "linear combination" of the solutions.

```
>  x1:=c1*sol1+c2*sol2+c3*sol3;
```

$$x1 := c1\,e^{(5\,t)} \begin{bmatrix} 2 \\ 1 \\ 0 \end{bmatrix} + c2\,e^{t} \begin{bmatrix} -1 \\ 0 \\ 1 \end{bmatrix} + c3\,e^{(-t)} \begin{bmatrix} 1 \\ 2 \\ -3 \end{bmatrix}$$

We need to find $c1, c2$, and $c3$ so that the initial condition is satisfied. We substitute $t = 0$ into the expression above and set it equal to the initial data vector. (**Caveat:** Remember that the subs command only works on a vector or a matrix that has been evaluated. Thus, subs(t=0,M); will *not* perform the substitution for any entries t in the array: evalm(M); will show no change.)

```
>  subs(t=0,map(evalm,x1=x0)):
```

```
>  ME:=map(simplify,lhs("))=rhs(");
```

$$ME := \begin{bmatrix} 2\,c1 - c2 + c3 \\ c1 + 2\,c3 \\ c2 - 3\,c3 \end{bmatrix} = \begin{bmatrix} 2 \\ 1 \\ -1 \end{bmatrix}$$

Maple's `equate` command in the `student` package will construct a set of equations from the preceding matrix equation *ME*.

```
> with(student):
```

```
> eqs:=equate(lhs(ME),rhs(ME));
```

$$eqs := \{c2 - 3\,c3 = -1,\ c1 + 2\,c3 = 1,\ 2\,c1 - c2 + c3 = 2\}$$

```
> csol:=solve(eqs,{c1,c2,c3});
```

$$csol := \{c1 = \frac{2}{3},\ c3 = \frac{1}{6},\ c2 = \frac{-1}{2}\}$$

An alternative approach is to use the fact that when Maple expects an equation and finds only an expression, it assumes that the equation is given by setting the expression equal to zero. Thus, when we use the `$` operator to form a sequence from the elements in the matrix difference, Maple views the elements of the matrix difference as equations.

```
> M:=evalm(lhs(ME)-rhs(ME));
```

$$M := \begin{bmatrix} 2\,c1 - c2 + c3 - 2 \\ c1 + 2\,c3 - 1 \\ c2 - 3\,c3 + 1 \end{bmatrix}$$

```
> sol:=solve({M[i,1]$i=1..3},{c1,c2,c3});
```

$$sol := \{c1 = \frac{2}{3},\ c3 = \frac{1}{6},\ c2 = \frac{-1}{2}\}$$

So our solution is given by

```
> x1soln:=subs(sol,evalm(x1));
```

$$x1soln := \begin{bmatrix} \frac{4}{3}\,e^{(5t)} + \frac{1}{2}\,e^{t} + \frac{1}{6}\,e^{(-t)} \\ \frac{2}{3}\,e^{(5t)} + \frac{1}{3}\,e^{(-t)} \\ -\frac{1}{2}\,e^{t} - \frac{1}{2}\,e^{(-t)} \end{bmatrix}$$

Let's check our work.

```
> map(diff,x1soln,t)-A1&*x1soln:
```

```
> evalm(");
```

$$\begin{bmatrix} 0 \\ 0 \\ 0 \end{bmatrix}$$

6.4.2 Repeated Eigenvalues with Insufficient Eigenvectors

Now consider the problem of trying to solve the initial value problem (6.1), (6.2) where $\mathbf{A} = \mathbf{A2}$ is given by

```
>  A2:=matrix(3,3,[-6,2,2,-13,-4,1,15,-5,-10]);
```

$$A2 := \begin{bmatrix} -6 & 2 & 2 \\ -13 & -4 & 1 \\ 15 & -5 & -10 \end{bmatrix}$$

and $\mathbf{x0}$ is the same as before. Again, we need the eigenvalues and eigenvectors of $\mathbf{A2}$.

```
>  eigenvals(A2);
```

$$-10, \ -5, \ -5$$

```
>  ev:=eigenvects(A2);
```

$$ev := [-5, \ 2, \ \{[0, \ -1, \ 1]\}], \ [-10, \ 1, \ \{[1, \ 3, \ -5]\}]$$

Here we have the trouble of dealing with a repeated eigenvalue that does not have enough eigenvectors. (The matrix is 3 by 3, but there are only two eigenvectors listed. One of the eigenvalues has multiplicity 2, but shows only one eigenvector.)

Now

$$s = \begin{bmatrix} 0 \\ 1 \\ -1 \end{bmatrix}$$

is an eigenvector for the eigenvalue -5. That means that $(A2 + 5Id)s = 0$. Since the eigenvalue -5 has fewer eigenvectors (1) shown than its multiplicity (2), we need to find a second vector, a so called "generalized eigenvector" for -5. The requirement for the new vector u is that $(A2 + 5Id)^2 u = 0$. This requirement is the same as saying that $(a2 + 5Id)u = s$. We solve for such vectors.

```
>  A2+5*diag(1,1,1);evalm(");
```

$$A2 + 5 \begin{bmatrix} 1 & 0 & 0 \\ 0 & 1 & 0 \\ 0 & 0 & 1 \end{bmatrix}$$

$$\begin{bmatrix} -1 & 2 & 2 \\ -13 & 1 & 1 \\ 15 & -5 & -5 \end{bmatrix}$$

```
>  uu:=linsolve(",vector(3,[0,-1,1]));
```

$$uu := \left[\frac{2}{25}, \ \_t_1, \ -\_t_1 + \frac{1}{25} \right]$$

Here $\_t_1$ can be chosen arbitrarily, say $\_t_1 = 0$.

```
>  subs(_t[1]=0,evalm(uu));
```

$$\left[\frac{2}{25},\, 0,\, \frac{1}{25}\right]$$

```
>  u:=convert(",matrix);
```

$$u := \begin{bmatrix} \dfrac{2}{25} \\[2mm] 0 \\[2mm] \dfrac{1}{25} \end{bmatrix}$$

We can then form three solutions.

```
>  exp(-5*t)*matrix(3,1,[0,-1,1]):
>  sol1:=evalm(");
```

$$sol1 := \begin{bmatrix} 0 \\ -e^{(-5\,t)} \\ e^{(-5\,t)} \end{bmatrix}$$

```
>  exp(-5*t)*(u+t*matrix(3,1,[0,-1,1])):
>  evalm("):
>  map(expand,"):
>  sol2:=map(simplify,",exp);
```

$$sol2 := \begin{bmatrix} \dfrac{2}{25}\,e^{(-5\,t)} \\[2mm] -e^{(-5\,t)}\,t \\[2mm] \dfrac{1}{25}\,e^{(-5\,t)} + e^{(-5\,t)}\,t \end{bmatrix}$$

```
>  exp(-10*t)*matrix(3,1,[1,3,-5]):
>  sol3:=evalm(");
```

$$sol3 := \begin{bmatrix} e^{(-10\,t)} \\ 3\,e^{(-10\,t)} \\ -5\,e^{(-10\,t)} \end{bmatrix}$$

We can check that the columns are solutions by augmenting them into a fundamental solution matrix **F**. We show that **F** satisfies the differential equation.

```
> augment(sol1,sol2,sol3):
```

```
> map(expand,"):
```

```
> F:=map(simplify,",exp);
```

$$F := \begin{bmatrix} 0 & \dfrac{2}{25}\,e^{(-5\,t)} & e^{(-10\,t)} \\[2ex] -e^{(-5\,t)} & -e^{(-5\,t)}\,t & 3\,e^{(-10\,t)} \\[2ex] e^{(-5\,t)} & \dfrac{1}{25}\,e^{(-5\,t)} + e^{(-5\,t)}\,t & -5\,e^{(-10\,t)} \end{bmatrix}$$

```
> map(diff,F,t)-A2&*F:evalm("):
```

```
> map(simplify,");
```

$$\begin{bmatrix} 0 & 0 & 0 \\ 0 & 0 & 0 \\ 0 & 0 & 0 \end{bmatrix}$$

We check that the solutions are independent by taking the Wronskian determinant and verifying that it is never zero.

```
> det(F):simplify(");
```

$$-\frac{1}{5}\,e^{(-20\,t)}$$

Our general solution will be a linear combination of these three. (Equivalently, we could have multiplied the fundamental matrix on the right by a column matrix of c's.)

```
> c1*sol1+c2*sol2+c3*sol3:evalm("):
```

```
> map(expand,"):
```

```
> x2:=map(simplify,",exp);
```

$$x2 := \begin{bmatrix} \dfrac{2}{25}\,c2\,e^{(-5\,t)} + c3\,e^{(-10\,t)} \\[2ex] -c1\,e^{(-5\,t)} - c2\,e^{(-5\,t)}\,t + 3\,c3\,e^{(-10\,t)} \\[2ex] c1\,e^{(-5\,t)} + \dfrac{1}{25}\,c2\,e^{(-5\,t)} + c2\,e^{(-5\,t)}\,t - 5\,c3\,e^{(-10\,t)} \end{bmatrix}$$

We need to use our initial data vector to determine $c1, c2, c3$.

```
>   subs(t=0,evalm(x2=x0)):
>   ME:=map(simplify,lhs("))=rhs(");
```

$$ME := \begin{bmatrix} \dfrac{2}{25}\, c2 + c3 \\[2mm] -c1 + 3\, c3 \\[2mm] c1 + \dfrac{1}{25}\, c2 - 5\, c3 \end{bmatrix} = \begin{bmatrix} 2 \\ 1 \\ -1 \end{bmatrix}$$

```
>   eqs:=equate(lhs(ME),rhs(ME));
```

$$eqs := \{-c1 + 3\, c3 = 1,\ c1 + \frac{1}{25}\, c2 - 5\, c3 = -1,\ \frac{2}{25}\, c2 + c3 = 2\}$$

```
>   csol:=solve(eqs,{c1,c2,c3});
```

$$csol := \{c2 = 20,\ c3 = \frac{2}{5},\ c1 = \frac{1}{5}\}$$

Therefore, our solution is given by

```
>   subs(csol,evalm(x2)):
>   x2soln:=map(simplify,",exp);
```

$$x2soln := \begin{bmatrix} \dfrac{8}{5}\, e^{(-5t)} + \dfrac{2}{5}\, e^{(-10t)} \\[3mm] -\dfrac{1}{5}\, e^{(-5t)} - 20\, e^{(-5t)}\, t + \dfrac{6}{5}\, e^{(-10t)} \\[3mm] e^{(-5t)} + 20\, e^{(-5t)}\, t - 2\, e^{(-10t)} \end{bmatrix}$$

```
>   map(diff,x2soln,t)-A2&*x2soln:evalm(");
```

$$\begin{bmatrix} 0 \\ 0 \\ 0 \end{bmatrix}$$

6.4.3 Complex Eigenvalues

Now let's consider an example with real coefficients but complex eigenvalues. Again we consider the problem of solving (6.1), (6.2) with $\mathbf{A} = \mathbf{A3}$ and $\mathbf{x0}$ as follows:

```
>   A3:=matrix(3,3,[5,-5,-5,-1,4,2,3,-5,-3]);
```

$$A3 := \begin{bmatrix} 5 & -5 & -5 \\ -1 & 4 & 2 \\ 3 & -5 & -3 \end{bmatrix}$$

```
>   x0:=matrix(3,1,[1,2,-1]);
```

$$x0 := \begin{bmatrix} 1 \\ 2 \\ -1 \end{bmatrix}$$

The eigenvalues and eigenvectors of **A3** are given by

```
>   ev:=eigenvects(A3);
```

$$ev := [2 + I, 1, \{[1, -\frac{2}{5} - \frac{1}{5}I, 1]\}], [2 - I, 1, \{[1, -\frac{2}{5} + \frac{1}{5}I, 1]\}], [2, 1, \{[0, -1, 1]\}]$$

Since the entries in the matrix are real numbers, both the eigenvalues and the eigenvectors come in complex conjugate pairs. (The complex conjugate of $a+bi$ is $a-bi$. For example, the eigenvalue $2+I$, is the conjugate of the eigenvalue $2 - I$.)

We follow the scheme of the first example, where the solutions have the form $ce^{\lambda t}\mathbf{v}$, where λ is an eigenvalue and $\mathbf{v}$ is an eigenvector associated with it. At first blush, it would appear that since λ is complex and $\mathbf{v}$ is complex, we would be doomed to having complex valued solutions. However, the constant c can *also* take on complex values, and the proper choices of c will yield linear combinations which are real-valued solutions. For each complex conjugate eigenvalue-eigenvector pair, there is a pair of independent real-valued solutions. These are the real and imaginary parts of $e^{\lambda t}\mathbf{v}$. Maple can compute them for us.

We assume as usual that t is a real-valued quantity, perhaps corresponding to time. To make this assumption explicit to Maple, we use the `assume` command. The fact that an assumption has been made on the variable t is indicated by a tilde placed near the t. (When we are past the part where we take real and imaginary parts, we can remove the assumption on t to improve the readability of the output.

```
>   assume(t,real):
```

```
>   lambda:=2+I;
```

$$\lambda := 2 + I$$

```
>   v:=matrix(3,1,[-2+I,1,-2+I]);
```

$$v := \begin{bmatrix} -2 + I \\ 1 \\ -2 + I \end{bmatrix}$$

```
>   exp(lambda*t)*v;
```

$$e^{((2+I)\,t^{\sim})}\,v$$

```
>   cxsol:=evalm(");
```

$$cxsol := \begin{bmatrix} (-2 + I)\,e^{((2+I)\,t^{\sim})} \\ e^{((2+I)\,t^{\sim})} \\ (-2 + I)\,e^{((2+I)\,t^{\sim})} \end{bmatrix}$$

```
> map(Re,evalm(cxsol)):

> sol1:=subs(t='t',evalm("));
```

$$sol1 := \begin{bmatrix} -2\,e^{(2\,t)}\cos(t) - e^{(2\,t)}\sin(t) \\ e^{(2\,t)}\cos(t) \\ -2\,e^{(2\,t)}\cos(t) - e^{(2\,t)}\sin(t) \end{bmatrix}$$

```
> map(Im,evalm(cxsol)):

> sol2:=subs(t='t',evalm("));
```

$$sol2 := \begin{bmatrix} -2\,e^{(2\,t)}\sin(t) + e^{(2\,t)}\cos(t) \\ e^{(2\,t)}\sin(t) \\ -2\,e^{(2\,t)}\sin(t) + e^{(2\,t)}\cos(t) \end{bmatrix}$$

The third solution is already real: (Note that after we have selected the real and imaginary parts, we no longer need the assumption on t, so we remove it.)

```
> exp(2*t)*matrix(3,1,[0,-1,1]):

> sol3:=subs(t='t',evalm("));
```

$$sol3 := \begin{bmatrix} 0 \\ -e^{(2\,t)} \\ e^{(2\,t)} \end{bmatrix}$$

```
> t:='t':
```

The general solution is a linear combination.

```
> c1*sol1+c2*sol2+c3*sol3:

> evalm("):

> x3:=map(simplify,evalm("));
```

$$x3 := \begin{bmatrix} -2\,c1\,e^{(2\,t)}\cos(t) - c1\,e^{(2\,t)}\sin(t) - 2\,c2\,e^{(2\,t)}\sin(t) + c2\,e^{(2\,t)}\cos(t) \\ c1\,e^{(2\,t)}\cos(t) + c2\,e^{(2\,t)}\sin(t) - c3\,e^{(2\,t)} \\ -2\,c1\,e^{(2\,t)}\cos(t) - c1\,e^{(2\,t)}\sin(t) - 2\,c2\,e^{(2\,t)}\sin(t) + c2\,e^{(2\,t)}\cos(t) + c3\,e^{(2\,t)} \end{bmatrix}$$

Let's get the values of $c1, c2$, and $c3$ from the initial data.

> subs(t=0,evalm(x3=x0)):

> ME:=map(simplify,lhs("))=rhs(");

$$ME := \begin{bmatrix} -2\,c1 + c2 \\ c1 - c3 \\ -2\,c1 + c2 + c3 \end{bmatrix} = \begin{bmatrix} 1 \\ 2 \\ -1 \end{bmatrix}$$

> eqns:=equate(lhs(ME),rhs(ME));

$$eqns := \{-2\,c1 + c2 = 1,\ c1 - c3 = 2,\ -2\,c1 + c2 + c3 = -1\}$$

> cts:=solve(eqns,{c1,c2,c3});

$$cts := \{c2 = 1,\ c3 = -2,\ c1 = 0\}$$

We find the particular solution *px3* that matches the initial conditions.

> px3:=subs(cts,evalm(x3));

$$px3 := \begin{bmatrix} -2\,e^{(2\,t)}\sin(t) + e^{(2\,t)}\cos(t) \\ e^{(2\,t)}\sin(t) + 2\,e^{(2\,t)} \\ -2\,e^{(2\,t)}\sin(t) + e^{(2\,t)}\cos(t) - 2\,e^{(2\,t)} \end{bmatrix}$$

Checking:

> map(diff,px3,t)-A3&*px3:evalm(");

$$\begin{bmatrix} 0 \\ 0 \\ 0 \end{bmatrix}$$

We remark in passing that the *easiest* way to get Maple to separate the real and imaginary solutions for us is to use the matrix exponential command that is discussed in the next section. Even when there are complex eigenvalues, the columns of exponential(A*t) are always real, linearly independent solutions.

> expAt:=exponential(A3*t):

> sol1:=matrix(3,1,col(expAt,1));

$$sol1 := \begin{bmatrix} 3\,e^{(2\,t)}\sin(t) + e^{(2\,t)}\cos(t) \\ -e^{(2\,t)}\sin(t) - e^{(2\,t)}\cos(t) + e^{(2\,t)} \\ 3\,e^{(2\,t)}\sin(t) + e^{(2\,t)}\cos(t) - e^{(2\,t)} \end{bmatrix}$$

```
> sol2:=matrix(3,1,col(expAt,2));
```

$$sol2 := \begin{bmatrix} -5\,e^{(2\,t)}\sin(t) \\ 2\,e^{(2\,t)}\sin(t) + e^{(2\,t)}\cos(t) \\ -5\,e^{(2\,t)}\sin(t) \end{bmatrix}$$

```
> sol3:=matrix(3,1,col(expAt,3));
```

$$sol3 := \begin{bmatrix} -5\,e^{(2\,t)}\sin(t) \\ e^{(2\,t)}\cos(t) - e^{(2\,t)} + 2\,e^{(2\,t)}\sin(t) \\ e^{(2\,t)} - 5\,e^{(2\,t)}\sin(t) \end{bmatrix}$$

Each of the columns of the resulting matrix is a real-valued solution to the differential equation, and the columns are independent.

SUMMARY: The eigenvalues of the matrix $\mathbf{A}$ determine the form of the solution of $\mathbf{x}'(t) = \mathbf{A}\mathbf{x}(t)$ in the following manner:

1. In our first example, the eigenvalues are $5, 1$, and -1, and the solution is made up of terms involving e^{5t}, e^t, and e^{-t}.

2. In our second example, the eigenvalues are given by $-10, -5$, and -5, and the solution is made up of terms containing e^{-10t}, e^{-5t}, and te^{-5t}. The lack of sufficient eigenvectors for the repeated eigenvalue -5 causes the term te^{-5t} to appear along with e^{-5t}.

3. In our last example, the eigenvalues are $2, 2 + I$, and $2 - I$; and the solution is made up of the terms involving $e^{2t}, e^{2t}\cos t$, and $e^{2t}\sin t$.

4. In general, a real eigenvalue a will give rise to terms of the form e^{at} if it is not repeated, and additional terms of the form $t^k e^{at}$ with k between 1 and $N - 1$ if it is repeated N times. Similarly, a complex eigenvalue $a \pm bI$ will give rise to terms of the form $e^{at}\cos(bt)$ and $e^{at}\sin(bt)$ if it is not repeated, and additional terms of the form $t^k e^{at}\cos(bt)$ and $t^k e^{at}\sin(bt)$ with k between 1 and $N - 1$ if it is repeated N times with only one eigenvector.

6.5   Nonhomogeneous First Order Linear Systems

This Maple section illustrates how the matrix exponential $e^{\mathbf{A}t}$ can be used to solve systems of the form

$$\mathbf{x}'(t) = \mathbf{A}\mathbf{x}(t) + \mathbf{f}(t), \quad \mathbf{x}(0) = \mathbf{x0}, \tag{6.3}$$

where $\mathbf{A}$ is an $n \times n$ matrix, $\mathbf{f}$ is a vector-valued function with n components, and $\mathbf{x0}$ is an n-vector. You might recall that the unique solution of (6.3) is given by

$$\mathbf{x}(t) = e^{\mathbf{A}t}\left[\mathbf{x0} + \int_0^t e^{-\mathbf{A}s}\,\mathbf{f}(s)\,ds\right]$$

Consequently, we can find the solution to our system if we can generate the matrix exponential and perform an integration.

Note that:

- If $\mathbf{f} = \mathbf{0}$, then the formula above gives the solution of the initial value problem

$$\mathbf{x}'(t) = \mathbf{A}\mathbf{x}(t), \quad \mathbf{x}(0) = \mathbf{x0}$$

 in the form

$$\mathbf{x}(t) = e^{\mathbf{A}t}\mathbf{x0}.$$

- The matrix $e^{-\mathbf{A}s}$ is the inverse of the matrix $e^{\mathbf{A}s}$.

Consider the example (6.3) with $\mathbf{A}, \mathbf{f}, \mathbf{x0}$ given by

```
>   A:=matrix(2,2,[7,-6,-6,-2]);
```

$$A := \begin{bmatrix} 7 & -6 \\ -6 & -2 \end{bmatrix}$$

```
>   f:=matrix(2,1,[1,sin(t)]);
```

$$f := \begin{bmatrix} 1 \\ \sin(t) \end{bmatrix}$$

```
>   x0:=matrix(2,1,[-1,1]);
```

$$x0 := \begin{bmatrix} -1 \\ 1 \end{bmatrix}$$

We will generate the matrix exponential $e^{\mathbf{A}t}$ and use the formula above to find our solution. Find the eigenvalues and eigenvectors of $\mathbf{A}$.

```
>   ev:=eigenvects(A);
```

$$ev := [-5, 1, \{[1, 2]\}], [10, 1, \{[-2, 1]\}]$$

Notice that our matrix has 2 distinct eigenvalues, there are enough eigenvectors. Form the matrix $\mathbf{C}$ by taking eigenvectors as columns. Next form a diagonal matrix using the exponentiated products of t and the eigenvalues.

```
>   C:=matrix(2,2,[1,-2,2,1]);
```

$$C := \begin{bmatrix} 1 & -2 \\ 2 & 1 \end{bmatrix}$$

```
>   expDt:=diag(exp(-5*t),exp(10*t));
```

$$expDt := \begin{bmatrix} e^{(-5\,t)} & 0 \\ 0 & e^{(10\,t)} \end{bmatrix}$$

The matrix exponential is the following product. That is, e^{At} is given by

```
>   expAt:=multiply(C,expDt,inverse(C));
```

$$expAt := \begin{bmatrix} \dfrac{1}{5}\,e^{(-5\,t)} + \dfrac{4}{5}\,e^{(10\,t)} & \dfrac{2}{5}\,e^{(-5\,t)} - \dfrac{2}{5}\,e^{(10\,t)} \\[2mm] \dfrac{2}{5}\,e^{(-5\,t)} - \dfrac{2}{5}\,e^{(10\,t)} & \dfrac{4}{5}\,e^{(-5\,t)} + \dfrac{1}{5}\,e^{(10\,t)} \end{bmatrix}$$

Notice that the columns in **C** are the eigenvectors of **A** listed *in the same order* in which the eigenvalues are used to form *expDt*. **This ordering is important!**

To solve our system, first form the integrand.

```
>   s:='s':exp_As:=subs(t=-s,evalm(expAt));
```

$$exp\_As := \begin{bmatrix} \dfrac{1}{5}\,e^{(5\,s)} + \dfrac{4}{5}\,e^{(-10\,s)} & \dfrac{2}{5}\,e^{(5\,s)} - \dfrac{2}{5}\,e^{(-10\,s)} \\[2mm] \dfrac{2}{5}\,e^{(5\,s)} - \dfrac{2}{5}\,e^{(-10\,s)} & \dfrac{4}{5}\,e^{(5\,s)} + \dfrac{1}{5}\,e^{(-10\,s)} \end{bmatrix}$$

```
>   fs:=subs(t=s,evalm(f));
```

$$fs := \begin{bmatrix} 1 \\ \sin(s) \end{bmatrix}$$

```
>   exp_As&*fs:map(expand,evalm(")):
>   intg:=map(simplify,evalm("),exp);
```

$$intg := \begin{bmatrix} \dfrac{1}{5}\,e^{(5\,s)} + \dfrac{4}{5}\,e^{(-10\,s)} + \dfrac{2}{5}\,\sin(s)\,e^{(5\,s)} - \dfrac{2}{5}\,\sin(s)\,e^{(-10\,s)} \\[2mm] \dfrac{2}{5}\,e^{(5\,s)} - \dfrac{2}{5}\,e^{(-10\,s)} + \dfrac{4}{5}\,\sin(s)\,e^{(5\,s)} + \dfrac{1}{5}\,\sin(s)\,e^{(-10\,s)} \end{bmatrix}$$

Now integrate.

```
>   map(int,intg,s=0..t):
>   map(expand,evalm(")):
>   int_port:=map(simplify,evalm("),exp);
```

$$int\_port :=$$
$$\left[\frac{1}{25}\,e^{(5\,t)} - \frac{2}{25}\,e^{(-10\,t)} - \frac{1}{65}\,e^{(5\,t)}\cos(t) + \frac{1}{13}\,\sin(t)\,e^{(5\,t)} + \frac{2}{505}\,e^{(-10\,t)}\cos(t) \right.$$
$$\left. + \frac{4}{101}\,\sin(t)\,e^{(-10\,t)} + \frac{1688}{32825} \right]$$
$$\left[\frac{2}{25}\,e^{(5\,t)} + \frac{1}{25}\,e^{(-10\,t)} - \frac{2}{65}\,e^{(5\,t)}\cos(t) + \frac{2}{13}\,\sin(t)\,e^{(5\,t)} - \frac{1}{505}\,e^{(-10\,t)}\cos(t) \right.$$
$$\left. - \frac{2}{101}\,\sin(t)\,e^{(-10\,t)} - \frac{2864}{32825} \right]$$

Our solution is given by

```
>  expAt&*(x0+int_port):
```

```
>  map(expand,evalm(")):
```

```
>  soln:=map(simplify,evalm("),exp);
```

$$soln := \begin{bmatrix} \dfrac{57}{325} e^{(-5t)} - \dfrac{1}{25} - \dfrac{15}{1313} \cos(t) + \dfrac{153}{1313} \sin(t) - \dfrac{2838}{2525} e^{(10t)} \\ \dfrac{114}{325} e^{(-5t)} + \dfrac{3}{25} - \dfrac{43}{1313} \cos(t) + \dfrac{176}{1313} \sin(t) + \dfrac{1419}{2525} e^{(10t)} \end{bmatrix}$$

Let's check to see whether this is actually the solution. We compute

$$\mathbf{x}'(t) - \mathbf{A}\mathbf{x}(t) - \mathbf{f}(t)$$

as follows:

```
>  evalm(map(diff,soln,t)-multiply(A,soln)-f);
```

$$\begin{bmatrix} 0 \\ 0 \end{bmatrix}$$

This tells us that the differential equation portion checks out. Now let's look at the initial data.

```
>  subs(t=0,evalm(soln)):
```

```
>  map(simplify,");
```

$$\begin{bmatrix} -1 \\ 1 \end{bmatrix}$$

This is **x0**, so *soln* is the solution of our initial value problem.

Maple can also form $e^{\mathbf{A}t}$ and $e^{-\mathbf{A}s}$ directly.

```
>  exponential(A*t);
```

$$\begin{bmatrix} \dfrac{1}{5} e^{(-5t)} + \dfrac{4}{5} e^{(10t)} & \dfrac{2}{5} e^{(-5t)} - \dfrac{2}{5} e^{(10t)} \\ \dfrac{2}{5} e^{(-5t)} - \dfrac{2}{5} e^{(10t)} & \dfrac{4}{5} e^{(-5t)} + \dfrac{1}{5} e^{(10t)} \end{bmatrix}$$

```
>  exponential(-A*s);
```

$$\begin{bmatrix} \dfrac{1}{5} e^{(5s)} + \dfrac{4}{5} e^{(-10s)} & \dfrac{2}{5} e^{(5s)} - \dfrac{2}{5} e^{(-10s)} \\ \dfrac{2}{5} e^{(5s)} - \dfrac{2}{5} e^{(-10s)} & \dfrac{4}{5} e^{(5s)} + \dfrac{1}{5} e^{(-10s)} \end{bmatrix}$$

Caveat: Note that there is a difference between the Maple commands `exp(A*t);`, which exponentiates each element in the matrix, and `exponential(A*t);`, which computes the *matrix exponential*, a very different object.)

6.6 Exercises

1. Use the `Help` facility in Maple to learn about the `dsolve` command. Then give the correct command to find the exact solution to the first order system

$$x'(t) = -2.3x(t) - 1.2y(t), \quad y'(t) = x(t) - 2.1y(t),$$

subject to the initial conditions $x(0) = 1$, $y(0) = -2$. Indicate the solution that Maple finds. (You may find it useful to turn off Maple's `label` feature with the command

`interface(labelling=false)`.)

2. Find $\lim_{t\to\infty} x(t)$ and $\lim_{t\to\infty} y(t)$ for the solutions $x(t)$ and $y(t)$ in Exercise 1.

3. Use `dsolve` to find the general solution to the first order system

$$x'(t) = y(t) + e^{-t}, \quad y'(t) = -x(t) - 2y(t) + 1.$$

4. Plot the direction field for the first order system given in Exercise 3 for $x(t)$ and $y(t)$ between -4 and 4. Then produce a plot of the direction field along with the parametric plots of solutions for each of the following initial conditions for t from 0 to 5.

$$x(0) = -1, \quad y(0) = 3;$$
$$x(0) = 2, \quad y(0) = -3;$$
$$x(1) = -1, \quad y(1) = 0.$$

5. Let a and b be arbitrary constants. Find a formula for the solution to the first order system

$$x'(t) = ax(t) - by(t), \quad y'(t) = bx(t) + ay(t),$$
$$x(0) = 1, \quad y(0) = 0.$$

Now sketch a coordinate system where the horizontal axis represents values of a and the vertical axis those of b. Which points in this coordinate system correspond to values of a and b for which the solution of the system will approach $(0,0)$ as $t \to \infty$?

6. Draw a picture which illustrates the flow paths of the mixing problem in Section 6.2.

7. Consider the mixing problem in Section 6.2 subject to the following changes. Suppose that at time $t = 0$ a mixture of 0.5 pounds per gallon begins flowing into tank 1 at the rate of r_1 gallons per minute, and mixed solution is pumped out at the same rate. In addition, solution is pumped into tank 2 (from tank 1) at the rate of r_2 gallons per minute, and mixed solution from tank 2 is returned to tank 1 at the same rate. Take all other information in the problem to be the same. Give the first order system which models this physical process. What will happen with the concentrations in the tanks as $t \to \infty$? Use both the form of the general solution to the first order system and a direction field plot to support your claim.

8. One of the simplest mathematical models for studying the spread of a disease is the so-called *SIR* model. The basic assumption is that the population is divided into three classes with respect to a disease. These classes are given by the *susceptible* class, the *infective* class and the *removed* class. The removed class consists of all individuals who have had the disease and have either died or recovered. If $S(t)$, $I(t)$,

and $R(t)$ respectively represent the number of people in each of the susceptible, infective, and removed classes at time t, then the standard *SIR* model is given by

$$
\begin{aligned}
S'(t) &= a - bS(t)I(t), \\
I'(t) &= bS(t)I(t) - cI(t), \\
R'(t) &= cI(t),
\end{aligned}
$$

where a, b, c are nonnegative constants. Here, a is considered to be a constant birth rate for the susceptible class. Explain what the model is saying about the rates at which each of the classes is changing. How do you think the model could be improved? Explain why we only need to study the first two equations in order to understand what happens to the population.

9. Consider the first two equations of the *SIR* model given in the Exercise 8:

$$
\begin{aligned}
S'(t) &= a - bS(t)I(t), \\
I'(t) &= bS(t)I(t) - cI(t).
\end{aligned}
$$

Take $b = 0.2$, $c = 0.1$, $S(0) = 100$, and $I(0) = 10$. Now choose several positive values for the parameter a and use Maple's DEplot command to analyze the behavior of solutions to the system. Use your plots to make a conjecture about the number of susceptibles and infectives in the population as $t \to \infty$.

10. Refer to the previous exercise, this time letting $a = 0$. Use DEplot to determine what happens to the susceptible and infective classes as $t \to \infty$. Does your answer change if the initial populations for the susceptibles and infectives change?

11. Use Maple to find the inverse of each of the following matrices.

$$
A = \begin{bmatrix} 1 & 2 \\ 3 & 4 \end{bmatrix} \quad
B = \begin{bmatrix} 1 & 2 & 3 \\ 3 & 4 & 5 \\ -5 & 6 & -7 \end{bmatrix} \quad
C = \begin{bmatrix} -2 & 1 & 0 & 0 \\ 1 & -2 & 1 & 0 \\ 0 & 1 & -2 & 1 \\ 0 & 0 & 1 & -2 \end{bmatrix}
$$

In each case, multiply the inverse matrix by the original matrix to verify Maple's computations.

12. Let a and b be real numbers. Find a formula for the inverse of the matrix

$$
\begin{bmatrix} a & -b \\ b & a \end{bmatrix}.
$$

Are there values of a and b for which you can not find the inverse?

13. Solve the system of equations

$$
\begin{aligned}
w - 2x - 3y + z &= -2, \\
3w + x - y + z &= 5, \\
2w - 3x + 7y - z &= 1, \\
-w + 2x - 3y + 2z &= -1.
\end{aligned}
$$

14. Find the eigenvalues and corresponding eigenvectors of these matrices.

$$\begin{bmatrix} 0 & 1 \\ -2 & -1 \end{bmatrix}, \quad \begin{bmatrix} -66 & -87 & -30 \\ 20 & 22 & 20 \\ 38 & 67 & 2 \end{bmatrix}, \quad \begin{bmatrix} 205 & 516 & -836 & 400 \\ -14 & -29 & 38 & -46 \\ 28 & 74 & -125 & 44 \\ -30 & -76 & 124 & -57 \end{bmatrix}.$$

15. Let a and b be real numbers. Find a formula for the eigenvalues and eigenvectors of the matrix

$$\begin{bmatrix} a & -b \\ b & a \end{bmatrix}.$$

16. This problem explores the relationship between the trace, determinant, and eigenvalues of a matrix.

   (a) Find a definition for the *trace* of an $n \times n$ matrix. (*Hint:* The `linalg[trace]` command is part of Maple's `linalg` package. Evaluate `trace(A)`; for three or four square matrices **A**.) What different object do you get if you look up `trace` in "Help" before reading in `with(linalg)`?

   (b) Randomly choose ten 2×2 matrices. (These are readily generated with the `randmatrix` command in `linalg`.) Find the determinant, the trace, and the eigenvalues of each of these matrices. (The determinant is computed with Maple's `det` command.) Can you see any relationship between these values? If not, then choose 10 more 2×2 matrices and try again!

   (c) When you think you know the relationship, choose 5 random 3×3 matrices and test your conjecture. Repeat your test with a random 4×4 matrix.

   (d) What do you think is true about the relationship between the trace, the determinant and the eigenvalues of *any* $n \times n$ matrix? (*Hint:* You might have to combine `evalf` and `Eigenvals` to test your conjecture.)

17. Suppose N is a given positive integer. Give a sequence of Maple commands which will create an $N \times N$ matrix whose (i, j) entry is given by $\dfrac{1}{i + j}$. Use these commands to create this matrix for each of $N = 2, 4, 6, 8, 10$. Find the determinant of each of these matrices.

18. Let **A** be the 10×10 matrix generated in the previous exercise. Suppose **b** is the column vector with 10 entries whose first entry is 20 and whose remaining entries are all 0. Find the solution to the system $\mathbf{Ax} = \mathbf{b}$. What do you think would happen if you used a calculator with 10 decimal places of accuracy to solve this problem?

In Exercises 19 through 23, find the solution to the linear first order system

$$\mathbf{x}'(t) = \mathbf{Ax}(t), \quad \mathbf{x}(0) = \mathbf{x0}$$

by using eigenvalues and eigenvectors. In each case, compare your solution with the one given by `dsolve`.

19. $\mathbf{A} = \begin{bmatrix} 0 & 1 \\ -2 & -1 \end{bmatrix}, \quad \mathbf{x0} = \begin{bmatrix} 1 \\ -2 \end{bmatrix}.$

20. $\mathbf{A} = \begin{bmatrix} -1 & 1 \\ -1 & -1 \end{bmatrix}, \quad \mathbf{x0} = \begin{bmatrix} 1 \\ -2 \end{bmatrix}.$

21. $\mathbf{A} = \begin{bmatrix} 1 & 3 & -9 \\ 1 & -1 & -3 \\ 1 & 1 & -5 \end{bmatrix}$, $\mathbf{x0} = \begin{bmatrix} -1 \\ 2 \\ 1 \end{bmatrix}$.

22. $\mathbf{A} = \begin{bmatrix} 13 & 21 & -57 \\ 19 & 33 & -89 \\ 10 & 17 & -46 \end{bmatrix}$, $\mathbf{x0} = \begin{bmatrix} -1 \\ 0 \\ 1 \end{bmatrix}$.

23. $\mathbf{A} = \begin{bmatrix} 13 & 20 & 46 \\ -28 & -37 & -80 \\ 8 & 10 & 21 \end{bmatrix}$, $\mathbf{x0} = \begin{bmatrix} -1 \\ 0 \\ 1 \end{bmatrix}$.

24. Let $A = \begin{bmatrix} -8 & 1 & -1 & 3 \\ 2 & -7 & 2 & 1 \\ 3 & 1 & -9 & 4 \\ 2 & 1 & 1 & -5 \end{bmatrix}$.

Use eigenvalues to predict the behavior as $t \to \infty$ of solutions to the first order system

$$\mathbf{x}'(t) = \mathbf{Ax}(t)$$

In exercises 25 through 29, find the solution to the linear first order system

$$\mathbf{x}'(t) = \mathbf{Ax}(t) + \mathbf{f}(t), \quad \mathbf{x}(0) = \mathbf{x0}$$

by finding the fundamental matrix for the system. In each case, compare your solution with the one given by dsolve.

25. $\mathbf{A} = \begin{bmatrix} 0 & 1 \\ -2 & -1 \end{bmatrix}$, $\mathbf{x0} = \begin{bmatrix} 1 \\ -2 \end{bmatrix}$, and $\mathbf{f}(t) = \begin{bmatrix} \sin t \\ 1 \end{bmatrix}$.

26. $\mathbf{A} = \begin{bmatrix} -1 & 1 \\ -1 & -1 \end{bmatrix}$, $\mathbf{x0} = \begin{bmatrix} 1 \\ -2 \end{bmatrix}$, and $\mathbf{f}(t) = \begin{bmatrix} 0 \\ \cos t \end{bmatrix}$.

27. $\mathbf{A} = \begin{bmatrix} 1 & 3 & -9 \\ 1 & -1 & -3 \\ 1 & 1 & -5 \end{bmatrix}$, $\mathbf{x0} = \begin{bmatrix} -1 \\ 2 \\ 1 \end{bmatrix}$, and $\mathbf{f}(t) = \begin{bmatrix} \sin t \\ \cos t \\ e^{-t} \end{bmatrix}$.

28. $\mathbf{A} = \begin{bmatrix} 13 & 21 & -57 \\ 19 & 33 & -89 \\ 10 & 17 & -46 \end{bmatrix}$, $\mathbf{x0} = \begin{bmatrix} -1 \\ 0 \\ 1 \end{bmatrix}$, and $\mathbf{f}(t) = \begin{bmatrix} 0 \\ 0 \\ 0 \end{bmatrix}$.

29. $\mathbf{A} = \begin{bmatrix} 13 & 20 & 46 \\ -28 & -37 & -80 \\ 8 & 10 & 21 \end{bmatrix}$, $\mathbf{x0} = \begin{bmatrix} -1 \\ 0 \\ 1 \end{bmatrix}$, and $\mathbf{f}(t) = \begin{bmatrix} 1 - \cos t \\ \sin 2t \\ 0 \end{bmatrix}$.

Chapter 7

The Phase Plane

It often possible to learn almost as much about the behavior of a system of equations from an analysis of its eigenvalues and eigenvectors as from an explicit formula for its solution. The methods are geometrical rather than analytic, and it is surprising how helpful a computer is in the geometric analysis. We begin with linear systems in two dimensions. Rather than plot solutions as a function of time, we plot the direction field and characteristic curves in the *phase plane*, where time does not appear explicitly.

7.1 Linear Systems

The general form of a linear two dimensional system with constant real coefficients is

$$\frac{d\mathbf{x}}{dt} = \mathbf{A}\mathbf{x},$$

where $\mathbf{A}$ is a 2×2 real matrix. The characteristic equation of $\mathbf{A}$ is quadratic. The analysis is based on the eigenvalues of $\mathbf{A}$.

Although these problems are easy in the sense that it is not difficult to write a formula for the general solution, it is often better to understand geometrically the behavior of solutions. Later, when nonlinear systems are studied, these same techniques will again prove to be helpful.

7.1.1 Real Eigenvalues

Distinct Negative Eigenvalues

```
> with(linalg):with(DEtools):

Warning, new definition for hilbert

Warning, new definition for norm

Warning, new definition for trace
```

```
>  diffeq1:=diff(y1(x),x)=-2*y1(x)+y2(x);
```

$$diffeq1 := \frac{\partial}{\partial x} y1(x) = -2\,y1(x) + y2(x)$$

```
>  diffeq2:=diff(y2(x),x)=y1(x)-2*y2(x);
```

$$diffeq2 := \frac{\partial}{\partial x} y2(x) = y1(x) - 2\,y2(x)$$

```
>  diffeqs:=diffeq1,diffeq2:
>  A:=matrix([[-2,1],[1,-2]]);eigenvects(A);
```

$$A := \begin{bmatrix} -2 & 1 \\ 1 & -2 \end{bmatrix}$$

$$[-1,\ 1,\ \{[1,\ 1]\}],\ [-3,\ 1,\ \{[-1,\ 1]\}]$$

```
>  inits:=[0,1,2],[0,-1,2],[0,2,-1],[0,-2,-1]:
>  show:=y1=-2..2,y2=-2..2:
>  T:=title='Distinct Negative Eigenvectors':
>  DEplot({diffeqs},[y1(x),y2(x)],x=-2..2,[inits],show,T);
```

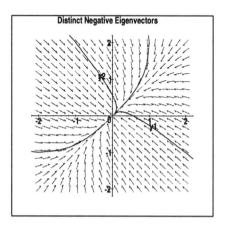

As you can see, the trajectories in the plot decay to (0,0) tangent to the eigenvector [1 1] belonging to -1, the largest eigenvalue. Can you think of the reason for this? (See Exercise 1.) The origin $\mathbf{x} = \mathbf{0}$ is an *attractor* because any solution with initial conditions "close" to the origin will be attracted to the origin. The origin is stable, since trajectories that begin near the origin will remain near the origin.

Distinct Positive Eigenvalues

```
>  diffeq1:=diff(y1(x),x)=2*y1(x)+y2(x);
```

$$diffeq1 := \frac{\partial}{\partial x} y1(x) = 2\,y1(x) + y2(x)$$

```
>  diffeq2:=diff(y2(x),x)=y1(x)+2*y2(x);
```

$$diffeq2 := \frac{\partial}{\partial x} y2(x) = y1(x) + 2\,y2(x)$$

```
>  diffeqs:=diffeq1,diffeq2:
>  A:=matrix(2,2,[2,1,1,2]);eigenvects(A);
```

$$A := \begin{bmatrix} 2 & 1 \\ 1 & 2 \end{bmatrix}$$

$$[3,\ 1,\ \{[1,\ 1]\}],\ [1,\ 1,\ \{[1,\ -1]\}]$$

```
>  inits:=[0,1,1.5],[0,-1,1.5],[0,1.5,-1],[0,-1.5,-1]:
>  show:=y1=-2..2,y2=-2..2:
>  T:=title='Distinct Positive Eigenvectors':
>  DEplot({diffeqs},[y1(x),y2(x)],x=-2..2,[inits],show,T);
```

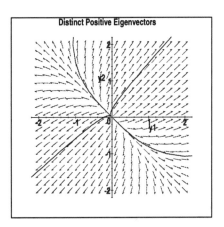

It is harder to see in this plot, but the trajectories $\mathbf{x}(t)$ tend to be parallel to the eigenvector $[1\ 1]$ belonging to the largest eigenvalue 3 as $t \to \infty$. (Again, see Exercise 1, as well as Exercise 2.)

The origin $\mathbf{x} = \mathbf{0}$ is a *repeller* because any solution with initial conditions "close" enough to the origin is repelled from the origin. The origin is therefore unstable.

Real Eigenvalues, Opposite Signs

In this example, the eigenvalues are the numbers 1 and −3:

>   `diffeq1:=diff(y1(x),x)=-y1(x)+2*y2(x);`

$$diffeq1 := \frac{\partial}{\partial x}\, y1(x) = -y1(x) + 2\, y2(x)$$

>   `diffeq2:=diff(y2(x),x)=2*y1(x)-y2(x);`

$$diffeq2 := \frac{\partial}{\partial x}\, y2(x) = 2\, y1(x) - y2(x)$$

>   `diffeqs:=diffeq1,diffeq2:`

>   `A:=matrix(2,2,[-1,2,2,-1]);eigenvects(A);`

$$A := \begin{bmatrix} -1 & 2 \\ 2 & -1 \end{bmatrix}$$

$$[1,\ 1,\ \{[1,\ 1]\}],\ [-3,\ 1,\ \{[-1,\ 1]\}]$$

>   `inits:=[0,1,1.5],[0,-1,1.5],[0,1.5,-1],[0,-1.5,-1],[0,0.5,-1]:`

>   `show:=y1=-2..2,y2=-2..2:`

>   `T:=title='Real Eigenvalues, Opposite Signs':`

>   `DEplot({diffeqs},[y1(x),y2(x)],x=-2..2,[inits],show,T);`

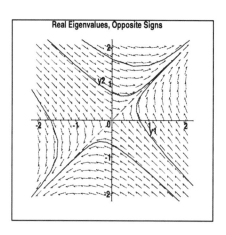

In this example the origin is called a *saddle point*; the node **x** = **0** is unstable.

Repeated Eigenvalue, Enough Eigenvectors

In this situation, a repeated eigenvalue must have *algebraic* multiplicity 2. The question remains whether there are one or two independent eigenvectors, that is, whether the *geometric* multiplicity of the eigenvalue is one or two. In either case, the node is unstable if the eigenvalue is positive and asymptotically stable if it is negative. If 0 is a repeated eigenvalue, then the situation depends on **A**. (See Exercise 3.) In any event, a 2×2 matrix with a double eigenvalue and enough eigenvectors is special. (See Exercise 4.)

```
> diffeq1:=diff(y1(x),x)=-y1(x);
```
$$diffeq1 := \frac{\partial}{\partial x} y1(x) = -y1(x)$$

```
> diffeq2:=diff(y2(x),x)=-y2(x);
```
$$diffeq2 := \frac{\partial}{\partial x} y2(x) = -y2(x)$$

```
> diffeqs:=diffeq1,diffeq2:
> A:=matrix(2,2,[-1,0,0,-1]);eigenvects(A);
```
$$A := \begin{bmatrix} -1 & 0 \\ 0 & -1 \end{bmatrix}$$

$$[-1,\ 2,\ \{[1,\ 0],\ [0,\ 1]\}]$$

```
> inits:=[0,1,1],[0,-1,1],[0,1,-1],[0,-1,-1]:
> show:=y1=-2..2,y2=-2..2:
> T:=title='Repeated Eigenvalue, Enough Eigenvectors':
> DEplot({diffeqs},[y1(x),y2(x)],x=-2..2,[inits],show,T);
```

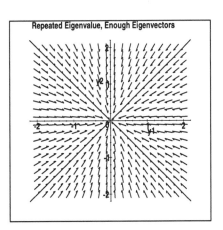

As an inspection of the plot suggests, the trajectories are all straight lines pointing toward the origin.

Repeated Eigenvalue, Insufficient Eigenvectors

In this case, which is typical for the 2×2 case, let **v** be an eigenvector belonging to the repeated root r; the general solution is

$$\mathbf{v} = c_1 \mathbf{v} e^{rt} + c_2 (\mathbf{v} t e^{rt} + \mathbf{u} e^{rt}),$$

where the vector **u** is a generalized eigenvector.

```
>  diffeq1:=diff(y1(x),x)=y1(x)+y2(x);
```

$$diffeq1 := \frac{\partial}{\partial x} y1(x) = y1(x) + y2(x)$$

```
>  diffeq2:=diff(y2(x),x)=y2(x);
```

$$diffeq2 := \frac{\partial}{\partial x} y2(x) = y2(x)$$

```
>  diffeqs:=diffeq1,diffeq2:
>  A:=matrix(2,2,[1,1,0,1]);eigenvects(A);
```

$$A := \begin{bmatrix} 1 & 1 \\ 0 & 1 \end{bmatrix}$$

$$[1, 2, \{[1, 0]\}]$$

```
>  inits:=[0,1,1],[0,-1,1],[0,1,-1],[0,-1,-1]:
>  show:=y1=-2..2,y2=-2..2:
>  T:=title='Repeated Eigenvalue, Insufficient Eigenvectors':
>  DEplot({diffeqs},[y1(x),y2(x)],x=-2..2,[inits],show,T);
```

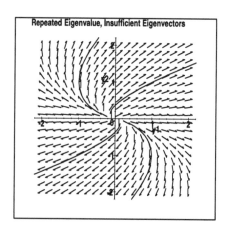

Note that the eigenvalue 1 only has one eigenvector [1 0]. The trajectories tend to the origin tangent to the x_1-axis as $t \to -\infty$ and away from the origin as t becomes large, since the eigenvalue is positive. The reverse would be true if the sign were reversed.

7.1.2 Complex Eigenvalues

Consider the following system, whose eigenvalues are complex conjugates.

```
>  diffeq1:=diff(y1(x),x)=y1(x)+y2(x);
```

$$diffeq1 := \frac{\partial}{\partial x} \, y1(x) = y1(x) + y2(x)$$

```
>  diffeq2:=diff(y2(x),x)=-y1(x);
```

$$diffeq2 := \frac{\partial}{\partial x} \, y2(x) = -y1(x)$$

```
>  diffeqs:=diffeq1,diffeq2:
>  A:=matrix(2,2,[1,1,-1,0]);eigenvects(A);
```

$$A := \left[\begin{array}{cc} 1 & 1 \\ -1 & 0 \end{array} \right]$$

$$[\frac{1}{2} - \frac{1}{2} I \sqrt{3}, \, 1, \, \{ \left[-\frac{1}{2} + \frac{1}{2} I \sqrt{3}, \, 1 \right] \}], [\frac{1}{2} + \frac{1}{2} I \sqrt{3}, \, 1, \, \{ \left[-\frac{1}{2} - \frac{1}{2} I \sqrt{3}, \, 1 \right] \}]$$

```
>  inits:=[0,2,0],[0,1,0],[0,-1,-1]:
>  show:=y1=-2..2,y2=-2..2:
>  T:=title='Repeated Eigenvalue, Insufficient Eigenvectors':
>  DEplot({diffeqs},[y1(x),y2(x)],x=-2..2,[inits],show,T);
```

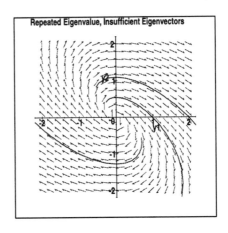

Here, the real part of the eigenvalue is positive, so the origin is a repeller. The trajectories appear to be exponential spirals. We may want to make another plot with a much greater range. As $t \to \infty$, do the trajectories continue to wind around the origin? (See Exercise 5.)

Purely Imaginary Eigenvalues

The following matrix has complex eigenvalues whose real parts are zero.

```
>  diffeq1:=diff(y1(x),x)=y2(x);
```

$$diffeq1 := \frac{\partial}{\partial x} y1(x) = y2(x)$$

```
>  diffeq2:=diff(y2(x),x)=-y1(x);
```

$$diffeq2 := \frac{\partial}{\partial x} y2(x) = -y1(x)$$

```
>  diffeqs:=diffeq1,diffeq2:
>  A:=matrix(2,2,[0,1,-1,0]);eigenvects(A);
```

$$A := \begin{bmatrix} 0 & 1 \\ -1 & 0 \end{bmatrix}$$

$$[I, 1, \{[-I, 1]\}], [-I, 1, \{[I, 1]\}]$$

```
>  inits:=[0,2,0],[0,1,0],[0,-1,-1]:
>  show:=y1=-2..2,y2=-2..2:
>  T:=title='Purely Imaginary Eigenvalues':
>  DEplot({diffeqs},[y1(x),y2(x)],x=-2..2,[inits],show,T);
```

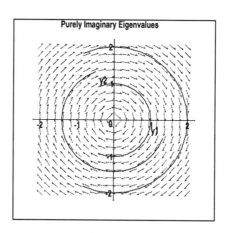

It is not hard to see that the trajectories are ellipses. (See Exercise 6.) The direction in which the trajectory is traversed depends on whether the imaginary part is positive or negative.

7.2 Autonomous Systems

In this section we apply the linear theory to certain nonlinear systems which have the form

$$\frac{dx}{dt} = f(x, y),$$

$$\frac{dy}{dt} = g(x, y),$$

where the functions f and g do not depend explicitly on time. The linear systems in the preceding section all had this form. What is different in this section is that f and g are no longer required to be linear.

A *critical point* (also called an *equilibrium point*) is a point (x_0, y_0) at which both $f(x_0, y_0) = 0$ and $g(x_0, y_0) = 0$. For a linear system this point is often only the origin (and always includes the origin). Unless **A** is singular, the origin is the only critical point of a linear system. Nonlinear systems can have other possibilities, as we will see.

For many interesting and important nonlinear systems we can associate with each critical point a linear system which will give us a qualitative understanding of the local behavior of the trajectories near that point. For this analysis, we need to know how to find the best linear approximation at a critical point. Linearization means replacing a nonlinear differential equation by a linear one which is in some sense close to the original equation, at least for some restricted range of the variables.

We will illustrate linearization for a system of two nonlinear, autonomous, first order differential equations. Consider the system

$$\frac{dx}{dt} = F(x, y), \quad \frac{dy}{dt} = G(x, y), \tag{7.1}$$

where we'll assume that the functions F and G are "nice;" i.e., they are differentiable functions with continuous partial derivatives. We also assume that at the point (x_0, y_0), both F and G are zero: $F(x_0, y_0) = G(x_0, y_0) = 0$. Such a point is called a *critical* point, or a *singular point*. It is easy to see that for such a point, the constant functions $x(t) = x_0$, $y(t) = y_0$ are solutions to (7.1). These are called *equilibrium* solutions. We mean by the term *linearization* of (7.1) at (x_0, y_0) the following system:

$$\frac{du}{dt} = F_x(x_0, y_0)u + F_y(x_0, y_0)v, \quad \frac{dv}{dt} = G_x(x_0, y_0)u + G_y(x_0, y_0)v, \tag{7.2}$$

or

$$\frac{d\mathbf{w}}{dt} = \mathbf{Aw}, \text{ where } \mathbf{A} = \begin{bmatrix} F_x & F_y \\ G_x & G_y \end{bmatrix} (x_0, y_0)$$

Equation (7.2) can be derived formally by replacing $F(x, y)$ and $G(x, y)$ by their first degree Taylor polynomials about (x_0, y_0), and then substituting u for $x - x_0$ and v for $y - y_0$. Maple has a command called `mtaylor` for computing such multidimensional Taylor polynomials. We illustrate this command below.

```
>   restart:with(DEtools):with(linalg):
```

```
Warning, new definition for norm
```

```
Warning, new definition for trace
```

```
>   readlib(mtaylor):F:='F':
```

```
>   Fapprox:=mtaylor(F(x,y),{x=x0,y=y0},2);
```

$$Fapprox := F(x0,\ y0) + D_1(F)(x0,\ y0)\,(x - x0) + D_2(F)(x0,\ y0)\,(y - y0)$$

```
>   Fapprox:=subs({F(x0,y0)=0,x=x0+u,y=y0+v},Fapprox);
```

$$Fapprox := D_1(F)(x0,\ y0)\,u + D_2(F)(x0,\ y0)\,v$$

The notation D_1 and D_2 is shorthand for $\dfrac{\partial}{\partial x}$ and $\dfrac{\partial}{\partial y}$, respectively.

It can be shown that in all but a few special cases, the behavior of the linearized solutions to (7.1) for (x, y) near (x_0, y_0) is essentially the same as that of solutions to (7.2) for (u, v) near $(0, 0)$. We will examine this equivalence in the case of a damped pendulum, which provides a nice example of an autonomous system. The differential equation that governs such a pendulum is

```
>   diff(theta(t),t$2)+1/2*diff(theta(t),t)+sin(theta(t))=0;
```

$$(\frac{\partial^2}{\partial t^2}\,\theta(t)) + \frac{1}{2}\,(\frac{\partial}{\partial t}\,\theta(t)) + \sin(\theta(t)) = 0$$

where θ is the angle to the downward vertical and t represents time.

If we let $x = \theta$ and $y = d\theta/dt$, we can write the second order pendulum equation as the first order system

$$\frac{dx}{dt} = F(x,y) = y, \quad \frac{dy}{dt} = G(x,y) = -\sin x - \frac{1}{2}y, \tag{7.3}$$

It is easily determined that the critical points for (7.3), where both derivatives are simultaneously zero, are $(x, y) = (n\pi, 0)$, where n can be any integer. We will use Maple to study the linearization of (7.3) near two of these, $(x_0, y_0) = (0, 0)$ and $(x_0, y_0) = (\pi, 0)$. We first make a plot showing the direction field for (7.3) and some typical solution trajectories.

```
>   F:=(x,y)->y:G:=(x,y)->-sin(x)-0.5*y:
```

```
>   init1:=[0,0,.5],[0,0,1],[0,0,1.5],[0,0,2],[0,1,2]:
```

```
>   init2:=[0,Pi-1.7,2],[0,Pi-1.5,2],[0,Pi-1,2]:
```

```
>   inits:=init1,init2:
```

```
>   diffeq1:=diff(x(t),t)=F(x(t),y(t));
```

$$diffeq1 := \frac{\partial}{\partial t}\,x(t) = y(t)$$

```
>   diffeq2:=diff(y(t),t)=G(x(t),y(t));
```

$$diffeq2 := \frac{\partial}{\partial t}\,y(t) = -\sin(x(t)) - .5\,y(t)$$

```
>   diffeqs:=diffeq1,diffeq2:

>   show:=x=-2..2+Pi,y=-2..2:

>   T:=title='Autonomous System':

>   DEplot([diffeqs],[x(t),y(t)],t=0..6,[inits],show,arrows=thin,T);
```

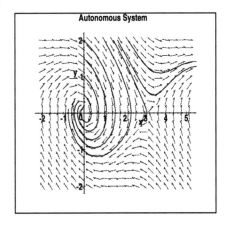

We next find the linearization of the system (7.3) about (0,0), and then make a plot of the phase plane for the associated linear system.

```
>   mtaylor(F(x,y),{x=0,y=0},2):

>   diffeq1:=diff(u(t),t)=subs({x=u(t),y=v(t)},");
```

$$diffeq1 := \frac{\partial}{\partial t}\, \mathrm{u}(t) = \mathrm{v}(t)$$

```
>   mtaylor(G(x,y),{x=0,y=0},2):

>   diffeq2:=diff(v(t),t)=subs({x=u(t),y=v(t)},");
```

$$diffeq2 := \frac{\partial}{\partial t}\, \mathrm{v}(t) = -1.\, \mathrm{u}(t) - .5\, \mathrm{v}(t)$$

```
>   diffeqs:=diffeq1,diffeq2:

>   inits:=[[0,0,.5],[0,0,1],[0,0,1.5],[0,0,2],[0,1,2]]:

>   show:=u=-2..2,v=-2..2:

>   T:=title='Critical Point [0,0]':

>   DEplot([diffeqs],[u(t),v(t)],t=0..6,inits,show,arrows=thin,T);
```

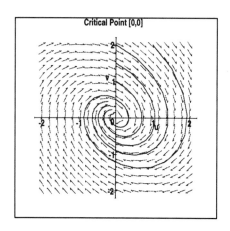

The similarity of the trajectories in this plot to trajectories in the original nonlinear system (7.3) is apparent. Next, we linearize about $(x_0, y_0) = (\pi, 0)$:

```
>  mtaylor(F(x,y),{x=Pi,y=0},2):
```

```
>  diffeq1:=diff(u(t),t)=subs({x=u(t)+Pi,y=v(t)},");
```

$$diffeq1 := \frac{\partial}{\partial t} u(t) = v(t)$$

```
>  mtaylor(G(x,y),{x=Pi,y=0},2):
```

```
>  diffeq2:=diff(v(t),t)=subs({x=u(t)+Pi,y=v(t)},");
```

$$diffeq2 := \frac{\partial}{\partial t} v(t) = u(t) - .5\, v(t)$$

```
>  diffeqs:=diffeq1,diffeq2:
```

```
>  inits:={[0,-2,2],[0,-1.7,2],[0,-1.5,2],[0,-1,2],[0,0,2]}:
```

```
>  show:=u=-2..2,v=-2..2:
```

```
>  T:=title='Critical Point [3.14,0]':
```

```
>  DEplot([diffeqs],[u,v],t=0..6,inits,show,arrows=thin,T);
```

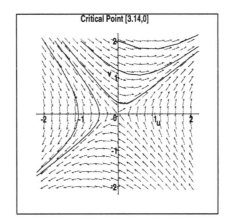

Again, the similarity to the nonlinear system is clear.

Moreover, now that we have the linear approximations, we recognize (7.3) the behaviors near each critical point. For example, near the first critical point, the origin, the matrix entries can be read from the coefficients in the linear differential equations:

```
>   A:=matrix(2,2,[0,1,-1,-0.5]);
```

$$A := \begin{bmatrix} 0 & 1 \\ -1 & -.5 \end{bmatrix}$$

The eigenvalues are

```
>   ev:=eigenvects(A);
```

$$ev := [-.2500000000 + .9682458366\,I,\ 1,\ \{[.9682458366 - .2500000000\,I,\ I]\}],$$
$$[-.2500000000 - .9682458366\,I,\ 1,\ \{[.9682458366 + .2500000000\,I,\ -I]\}]$$

and the critical point is seen to be an asymptotically stable attractor. This agrees with our intuition that a downward hanging pendulum will tend to return to a downward position if moved slightly off its equilibrium position.

On the other hand, near the second critical point at $[\pi, 0]$, the entries in the matrix are

```
>   A:=matrix(2,2,[0,1,1,-0.5]);
```

$$A := \begin{bmatrix} 0 & 1 \\ 1 & -.5 \end{bmatrix}$$

so the eigenvalues are

```
>   eigenvects(A);
```

$$[-1.280776406,\ 1,\ \{[.6154122097,\ -.7882054382]\}],$$
$$[.7807764067,\ 1,\ \{[-.7882054382,\ -.6154122097]\}]$$

and the critical point is an unstable saddle point. The model also agrees with our intuition that a pendulum raised straight up (angle π) will not remain near its original position if moved slightly off the equilibrium position.

Competing Species

In the following example, we will study a nonlinear autonomous system which is a mathematical model of interacting populations. In southeast Texas, the coyote *Canis latrans* (now more frequently a cross between the wild coyote and the domestic dog) and the raccoon *Bassaricus astutus* compete for prey.

A natural question arises: How can predators coexist? We know by this example that nature has solved this problem. Can we find a mathematical model of the competition?

Suppose a predator faces a limited food supply, which in a simple growth model can translate into the logistic equation

$$\frac{dx}{dt} = x(c_1 - s_1 x).$$

With two predators and with no interaction at all (imagine they pursued different prey), we add another logistic equation for the second predator y. If, however, they compete for the same prey, then the growth rates, which depend on the food supply, interact. Notice that this interaction is *not* due to the predators directly interacting, only to their seeking the same or similar prey.

The stress on the food supply can be modeled by adding a term $-axy$ to each of the competing species' logistic equations. The coefficient a is a measure of the degree to which each species interferes with the other. (Clearly, the value of a in the logistic equation does not have to be the same for both competing predators.)

For example, if the original logistic equations are $dx/dt = x(1 - x)$ and $dy/dt = y(0.75 - 0.5y)$, then using $a = 1$ in the first equation, for the effect of y on x, and $a = 0.5$ in the second equation, for the effect of x on y, we get

$$\frac{dx}{dt} = x(1 - x - y),$$
$$\frac{dy}{dt} = y(0.75 - y - 0.5x).$$

To find the critical points, where both derivatives are simultaneously zero, we look on the plot below. The derivative in the first equation is zero on the y-axis, where $x = 0$, and on the line $x + y = 1$. Similarly, the derivative in the second equation is zero on the x-axis and the line $y = 3/4 - x/2$.

```
> T:=title='Equilibrium Point':

> plot({1-x,3/4-x/2},x=0..3/2,0..3/2,T);
```

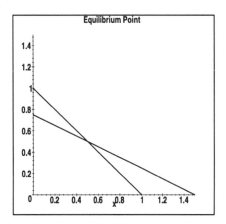

Notice there are four critical points: $(0,0)$, $(0,0.75)$, $(1,0)$, and $(0.5,0.5)$. For this model, the critical points with one or more of the species absent are not interesting.

Near the other critical point, $(0.5, 0.5)$, we can gain some insight into how the predators relate by examining how the trajectories behave.

```
>  diffeq1:=diff(y1(x),x)=y1(x)*(1-y1(x)-y2(x));
```

$$diffeq1 := \frac{\partial}{\partial x}\, y1(x) = y1(x)\,(1 - y1(x) - y2(x))$$

```
>  diffeq2:=diff(y2(x),x)=y2(x)*(0.75-y2(x)-0.5*y1(x));
```

$$diffeq2 := \frac{\partial}{\partial x}\, y2(x) = y2(x)\,(.75 - y2(x) - .5\,y1(x))$$

```
>  diffeqs:=diffeq1,diffeq2:

>  init1:=[0,1.5,1],[0,1.5,0.4],[0,1.5,0.7]:

>  init2:=[0,1.5,1.5],[0,0.1,0.01],[0,0.5,0.1]:

>  init3:=[0,0.1,0.6],[0,0.1,0.8],[0,0.1,1.1]:

>  inits:=init1,init2,init3:

>  show:=y1=0..1,y2=0..1:

>  T:=title='Competing Species':

>  DEplot({diffeqs},[y1(x),y2(x)],x=-3..3,[inits],show,T);
```

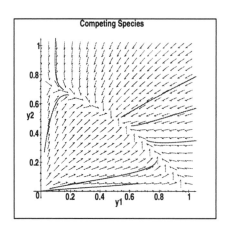

Notice that the trajectories appear to tend to the critical point, so that the predators in this model can coexist provided they begin close enough to this stable critical point.

We would have reached the same conclusion by the technique of linearization.

The linearization of the previous system at the point $(0.5, 0.5)$ is:

$$\frac{dx}{dt} = -\frac{1}{2}x - \frac{1}{2}y + \frac{1}{2}$$
$$\frac{dy}{dt} = -\frac{1}{4}x - \frac{1}{2}y + \frac{3}{8}$$

> diffeq1:=diff(y1(x),x)=-1/2*y1(x)-1/2*y2(x)+1/2;

$$diffeq1 := \frac{\partial}{\partial x}\, y1(x) = -\frac{1}{2}\, y1(x) - \frac{1}{2}\, y2(x) + \frac{1}{2}$$

> diffeq2:=diff(y2(x),x)=-1/4*y1(x)-1/2*y2(x)+3/8;

$$diffeq2 := \frac{\partial}{\partial x}\, y2(x) = -\frac{1}{4}\, y1(x) - \frac{1}{2}\, y2(x) + \frac{3}{8}$$

> diffeqs:=diffeq1,diffeq2:

> A:=matrix(2,2,[-1/2,-1/2,-1/4,-1/2]);eigenvects(A);

$$A := \begin{bmatrix} \dfrac{-1}{2} & \dfrac{-1}{2} \\ \dfrac{-1}{4} & \dfrac{-1}{2} \end{bmatrix}$$

$$\left[-\frac{1}{2} + \frac{1}{4}\sqrt{2},\, 1,\, \left\{\left[1,\, -\frac{1}{2}\sqrt{2}\right]\right\}\right],\, \left[-\frac{1}{2} - \frac{1}{4}\sqrt{2},\, 1,\, \left\{\left[1,\, \frac{1}{2}\sqrt{2}\right]\right\}\right]$$

```
>   init1:=[0,1.5,1],[0,1.5,0.4],[0,1.5,0.7]:

>   init2:=[0,1.5,1.5],[0,0.1,0.01],[0,0.5,0.1]:

>   init3:=[0,0.1,0.6],[0,0.1,0.8],[0,0.1,1.1]:

>   inits:=init1,init2,init3:

>   show:=y1=0..1,y2=0..1:

>   T:=title='Linearized Competing Species':

>   DEplot({diffeqs},[y1(x),y2(x)],x=-3..3,[inits],show,T);
```

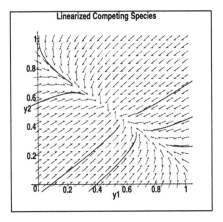

Notice that the trajectories are similar to those of the nonlinear system. The eigenvalues are $(-2 \pm \sqrt{2})/4$, which are both negative; so the equilibrium is stable.

Contrast the above situation with the following example which has only slightly different logistic equation (slightly different values of a), but the same critical point $(0.5, 0.5)$.

The system is

$$\frac{dx}{dt} = x(1 - x - y),$$
$$\frac{dy}{dt} = y(1/2 - y/4 - 3x/4).$$

To find the critical points, where both derivatives are simultaneously zero, we look on the plot below. The derivative in the first equation is zero on the y-axis, where $x = 0$, and on the line $x + y = 1$. Similarly, the derivative in the second equation is zero on the x-axis and on the line $y = 2 - 3x$.

```
>   T:=title='Equilibrium Point':

>   plot({1-x,2-3*x},x=0..3/2,0..3/2,T);
```

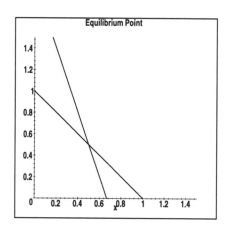

We are interested in the behavior near the point $[0.5, 0.5]$.

```
> diffeq1:=diff(x(t),t)=x(t)*(1-x(t)-y(t));
```

$$diffeq1 := \frac{\partial}{\partial t} x(t) = x(t)\,(1 - x(t) - y(t))$$

```
> diffeq2:=diff(y(t),t)=y(t)*(1/2-y(t)/4-3*x(t)/4);
```

$$diffeq2 := \frac{\partial}{\partial t} y(t) = y(t)\,(\frac{1}{2} - \frac{1}{4} y(t) - \frac{3}{4} x(t))$$

```
> diffeqs:=diffeq1,diffeq2:
```

```
> init1:=[0,1.1,0.5],[0,0.2,0.4],[0,0.3,0.3],[0,0.4,0.1]:
```

```
> init2:=[0,1,0.2],[0,1,0.4],[0,1,0.6],[0,1,0.8],[0,1,1]:
```

```
> init3:=[0,1,1.3],[0,1,1.6],[0,1,1.9]:
```

```
> inits:=init1,init2,init3:
```

```
> show:=x=0..1,y=0..1:
```

```
> T:=title='New Competing Species':
```

```
> DEplot({diffeqs},[x(t),y(t)],t=-3..3,[inits],show,T);
```

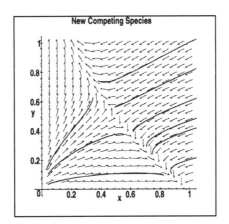

Here, the linearization of the system (in $(u, v) = (x - 0.5, y - 0.5)$) is

$$\frac{du}{dt} = -0.5u - 0.5v$$

$$\frac{dv}{dt} = -0.375u - 0.125v$$

which has eigenvalues $(-5 \pm \sqrt{57})/16$, of opposite signs, so that the critical point is a saddle point. This is seen in the unstable equilibrium of the direction field plot.

To derive the linear approximation at $[1/2, 1/2]$, we take the linear part of

```
>   subs({x=u+1/2,y=v+1/2},x*(1-x-y)):expand(");
```

$$-u^2 - uv - \frac{1}{2}u - \frac{1}{2}v$$

and the linear part of

```
>   subs({x=u+1/2,y=v+1/2},y*(1/2-y/4-3*x/4)):expand(");
```

$$-\frac{1}{4}v^2 - \frac{3}{4}uv - \frac{1}{8}v - \frac{3}{8}u$$

Thus, the matrix $\mathbf{A}$ is given by

```
>   A:=matrix(2,2,[-1/2,-1/2,-3/8,-1/8]);
```

$$A := \begin{bmatrix} \dfrac{-1}{2} & \dfrac{-1}{2} \\ \dfrac{-3}{8} & \dfrac{-1}{8} \end{bmatrix}$$

and its eigenvalues and eigenvectors by

```
>   ev:=eigenvects(A);
```

$$ev := [-\frac{5}{16} + \frac{1}{16}\sqrt{57}, 1, \{[1, -\frac{3}{8} - \frac{1}{8}\sqrt{57}]\}], [-\frac{5}{16} - \frac{1}{16}\sqrt{57}, 1, \{[1, -\frac{3}{8} + \frac{1}{8}\sqrt{57}]\}]$$

The Predator-Prey Equations

Another autonomous system comes from a mathematical model of a predator-prey situation. Assume that there is a predator species, the fox for example, that preys on rabbits, who in turn have an unlimited food supply.

Without foxes, the rabbits grow exponentially: $dr/dt = ar$. Without rabbits, the foxes decay exponentially: $df/dt = -cf$. With both species present, there is interaction which benefits foxes and diminishes rabbits.

The simplest model has the interaction term proportional to the product of the number of rabbits and the number of foxes. (If there are twice as many foxes and three times as many rabbits next year, there will be six times as many encounters). We denote by $x(t)$ number of the prey and by $y(t)$ the number of the predator.

```
>  diffeq1:=diff(x(t),t)=x(t)*(1-(1/2)*y(t));
```

$$diffeq1 := \frac{\partial}{\partial t} x(t) = x(t)\left(1 - \frac{1}{2} y(t)\right)$$

```
>  diffeq2:=diff(y(t),t)=y(t)*(-3/4+x(t)/4);
```

$$diffeq2 := \frac{\partial}{\partial t} y(t) = y(t)\left(-\frac{3}{4} + \frac{1}{4} x(t)\right)$$

```
>  diffeqs:=diffeq1,diffeq2:
```

The coefficients of the $x(t)y(t)$ terms represent the net effect on survival of a rabbit-fox encounter: the rabbit is not always caught at every encounter, and eating one rabbit does not in itself guarantee that a fox will survive; but the net effect is negative for the rabbits and positive for the foxes.

```
>  init1:=[0,3,.01],[0,3,.2],[0,3,.5],[0,3,.8]:

>  init2:=[0,3,1.1],[0,3,1.5],[0,3,1.8]:

>  inits:=init1,init2:

>  show:=x=0..5,y=0..5:

>  T:=title='Preditor-Prey':

>  DEplot({diffeqs},[x(t),y(t)],t=-3..3,[inits],show,T);
```

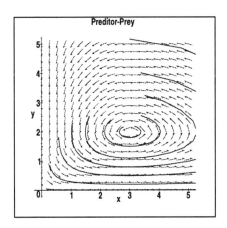

The point $(3, 2)$ appears to be an equilibrium around which there is a cyclic rise and fall of both populations. The result indicated in this model agrees with what one might expect: if the rabbit population increases, then the population of foxes will increase, but slightly later; and then over hunting will drive down the population of the rabbits, and slightly later, the foxes. (See also Exercise 8.)

7.3   Exercises

1. Let
$$x' = Ax,$$
   where A is a 2×2 constant real matrix having distinct real eigenvalues $r_1 > r_2$ and eigenvectors v_1, v_2 with
$$Av_k = r_k v_k, \quad k = 1, 2.$$
   Show that the general solution
$$x(t) = c_1 v_1 e^{r_1 t} + c_2 v_2 e^{r_2 t}$$
   satisfies
$$\lim_{t \to \infty} e^{-r_1 t} x(t) = c_1 v_1.$$

2. Replot the second example in Section 7.1 using the same initial conditions but with ranges -99..99, -99..99. Explain what you see in terms of the eigenvector of A belonging to the largest eigenvalue.

3. Suppose $A \neq 0$ is a 2×2 matrix which has 0 as a repeated eigenvalue. Show that 0 has geometric multiplicity 1. Characterize the solutions to $x' = Ax$. (*Hint:* First determine the conditions on A in order for it to have 0 as a repeated eigenvalue.)

4. Suppose A has a double eigenvalue. Show that, if there are enough eigenvectors, then A is a diagonal matrix. (You should not expect to generalize this special case to more than 2×2 matrices. It may help to remember material relating to diagonalizing matrices.)

5. Consider the first example in Subsection 7.1.2:

$$\mathbf{A} = \begin{bmatrix} 1 & 1 \\ -1 & 0 \end{bmatrix},$$

with eigenvalues $\dfrac{1 \pm \sqrt{3}\,i}{2}$. Let $\mathbf{x}(t)$ be any trajectory. Show that

$$e^{-\frac{1}{2}t}\,\mathbf{x}(t)$$

winds clockwise around the origin as $t \to \infty$.

6. In the system

$$\mathbf{x}' = \mathbf{A}\mathbf{x},$$

where $\mathbf{A}$ is a constant 2×2 matrix, show the eigenvalues are purely imaginary if and only if det $\mathbf{A} >$ 0 and tr $\mathbf{A} = 0$. (The *trace*, tr $\mathbf{A}$, of a matrix $\mathbf{A}$ is the sum of its diagonal entries). Show that the trajectories are ellipses.

7. Consider the autonomous system

$$\begin{aligned} \frac{dx}{dt} &= f(x, y), \\ \frac{dy}{dt} &= g(x, y), \end{aligned}$$

and let $(x(t), y(t))$ be a solution. Let c be fixed and define $u(t) = x(t + c)$ and $v(t) = y(t + c)$. Show that $(u(t), v(t))$ is a solution. (*Hint:* Use the chain rule.)

8. Consider the predator-prey problem discussed last:

$$\begin{aligned} \frac{df}{dt} &= f(1 - 0.5r) \\ \frac{dr}{dt} &= r(-.75 + 0.25f) \end{aligned}$$

Linearize this problem about the critical point $(f, r) = (3, 2)$. What are the eigenvalues? Classify the critical point of the linearized system.

9. Consider the nonlinear system

$$\begin{aligned} \frac{dx}{dt} &= -y + (1 - x^2 - y^2)x \\ \frac{dy}{dt} &= x + (1 - x^2 - y^2)y \end{aligned}$$

Show that $x = \cos t$, $y = \sin t$ is a solution, so that the unit circle is a trajectory. Show that the origin is the only critical point, and use Maple to illustrate that trajectories spiral toward the unit circle from inside and outside.

Chapter 8

Numerical Solutions of Systems

In the not too distant past, the traditional first course in differential equations (which is the only one taken by most students) paid little or no attention to the numerical solution of differential equations. This situation can at best be described as unfortunate, because **most differential equations arising from applications can only be solved numerically**.

In this chapter, we demonstrate how initial value problems can be analyzed numerically, first for individual equations, then for systems. Maple procedures will be built for Euler, modified Euler, and fourth order Runge-Kutta methods. We will consider the rates of convergence of these various methods, and alternatives to them like the Taylor series method. We will also show an example of how other numeric methods are applied to boundary value problems.

Next, we write an explicit procedure to implement Euler's method for a system, just to see what one looks like. Then we demonstrate the `numeric` option of `dsolve` for the same system. We then apply the `dsolve` command to do an example of the two dimensional motion of a projectile with friction. Since higher order equations can be reduced to systems of first order equations, we will then be able to use numerical methods for higher order equations.

Earlier, in Section 1.3, we first constructed a procedure for Euler's method. For completeness, we reproduce it here.

```
>   MyEuler:=proc(f,firstx,lastx,firsty,stepsize)
    local x,y,h,point,eulerseq,tempy;
    global eulerlist;
    y:=firsty;h:=stepsize;
    eulerseq:=NULL;
    for x from firstx by h to lastx do
    point:=[x,y];
    eulerseq:=eulerseq,point:tempy:=y;
    y:= evalhf(y+h*f(x,y));
    od;
    eulerlist:=[eulerseq];
    tempy;
    end:
```

In Euler's method the derivative is replaced by the standard difference quotient. This idea leads to a numerical scheme that has an intrinsic error of order $|h|$, where h is the step size. If the derivative is replaced with a more accurate quotient, we should get a better approximation to the solution to the initial value problem.

One choice of a more accurate approximation to the derivative is the symmetric difference quotient:

$$\frac{dy}{dx} \approx \frac{y(x+h) - y(x-h)}{2h}. \tag{8.1}$$

This leads to the recursive scheme

$$y_{k+1} = y_{k-1} + 2hf(x_k, y_k)$$

The last expression is somewhat more difficult to compute than the expression used in the old Euler's method, because the new value of y depends on the last *two* values of y. At the start of the algorithm, when we have only one value of y, we use the old Euler method to compute the second value of y in the first iteration. Procedures such as this, which require more than one previously computed value of y to compute a new value, are called *multistep methods*.

```
>  MyModEuler:=proc(f,firstx,lastx,firsty,stepsize)
   local x,y,h,ty,yy,point,points,modeulerseq;
   global modeulerlist;
   h:=stepsize;y:=firsty;
   yy:=MyEuler(f,firstx,firstx+stepsize,firsty,stepsize);
   points:=[firstx,firsty],[firstx+h,yy];modeulerseq:=points;
   for x from firstx+2*h by h to lastx do
   ty:=yy;yy:=evalhf(y+2*h*(f(x-h,yy)));y:=ty;point:=[x,yy];
   modeulerseq:=modeulerseq,point;
   od;
   modeulerlist:=[modeulerseq];
   yy;
   end:
```

Notice we have again used Maple's `evalhf` command.

We now use this modified Euler method on the initial value problem with *lastx* again equal to 10.

```
>  MyModEuler(f,0,10,1,0.1);
```

$$50.46260578995661$$

Notice that the result from our Euler's method is within 0.5 of the value computed by this modified Euler's method.

```
>  euler2:=plot(modeulerlist,0..10,0..55,linestyle=2):
```

```
>  display(euler2);
```

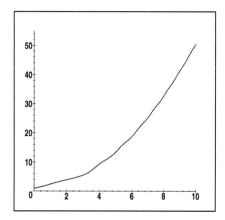

The next algorithm we implement with Maple is a Runge-Kutta method.

```
>   RK:= proc(f, firstx, lastx, firsty, stepsize)
    local h,x,y,i,k1,k2,k3,k4,point,rkseq;
    global rklist;
    h:=stepsize;x:=firstx;y:=firsty;point:=[x,y];rkseq:=point;
    while x<=lastx-h do
    k1:=evalhf(f(x, y));
    k2:=evalhf(f(x+0.5*h,y+0.5*h*k1));
    k3:=evalhf(f(x+0.5*h,y+0.5*h*k2));
    k4:=evalhf(f(x+h,y+h*k3));
    y:= evalhf(y+(h/6)*(k1+2*k2+2*k3+k4));
    x:=x+h;
    point:=[x,y];
    rkseq:=rkseq,point;
    od;
    rklist:=[rkseq];
    y;
    end:
```

Using this algorithm we compute yet another approximate value of the solution to our initial value problem at $x = 10$.

```
>   RK(f,0,10,1,0.1);
```

$$50.51795879796921$$

So far the three approximations we have computed for the value of the solution at the point $x = 10$ are approximately 50.085, 50.463, and 50.518. Which is closest to the actual value of the solution? One of the exercises at the end of this chapter asks you to try these three routines with the step size set to 0.05.

```
>   rk1:=plot(rklist,0..10,0..55,linestyle=2):
```

```
>   display(rk1);
```

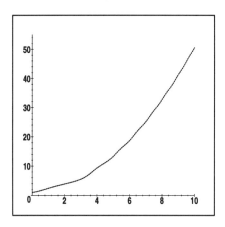

In Release 4, one can also implement classical methods such as Euler, Heun, and Runge-Kutta through the dsolve/numeric command. For example, we re-execute the Runge-Kutta approximation above.

> diffeq:=diff(y(x),x)=x+sin(y(x));

$$diffeq := \frac{\partial}{\partial x}\, y(x) = x + \sin(y(x))$$

> inits:=y(0)=1:

> myoptions:=method=classical[rk4],stepsize=0.1:

> sol:=dsolve({diffeq,inits},y(x),numeric,myoptions);

$$sol := \mathbf{proc}(x\_classical)\ \ldots\ \mathbf{end}$$

> sol(10);

$$[x = 10,\ y(x) = 50.51795879796934]$$

As a last attempt to find a good approximate value of the solution to this initial value problem, we use Maple's dsolve command with the numeric option.

> sol:=dsolve({diff(y(x),x)=f(x,y(x)),inits},y(x),numeric);

$$sol := \mathbf{proc}(rkf45\_x)\ \ldots\ \mathbf{end}$$

> sol(10);

$$[x = 10,\ y(x) = 50.51826211510741]$$

We now have several different values that approximate the solution to our initial value problem at $x = 10$. Which do you think is closest to the actual value?

8.1   The Order of Convergence for a Numerical Method

Maple will be used to illustrate the order of convergence; i.e., how fast the error goes to zero as the step size is decreased. For this purpose, we will use an initial value problem whose exact solution we can find.

```
>   diffeq:=diff(y(t),t)=(y(t)^2-y(t)/t)/2;inits:=y(1)=-1:
```

$$diffeq := \frac{\partial}{\partial t}\, y(t) = \frac{1}{2}\, y(t)^2 - \frac{1}{2}\, \frac{y(t)}{t}$$

```
>   sol:=dsolve({diffeq,inits},y(t));
```

$$sol := y(t) = -\frac{1}{t}$$

We will use the improved Euler's method (Heun method), a so called *second order* method. What does order of convergence mean and why is it useful to know the order of convergence of a numerical method? We will use Maple to illustrate this concept. A method is of order p if p is the largest positive integer such that the error, at a fixed value of t, is at most a constant times h^p, where h is the step size. Moreover, it is usually the case that the error is approximately equal to Ch^p, as $h \to 0$. In other words, the error divided by h^p is nearly constant for all small values of h.

To demonstrate this, we will implement this method to find approximations to the solution $y = -1/t$ at the point $t = 2$. The plan is to use several decreasing values $(2^{-k}, k = 1, 2, ..., 8)$, for the step size. Since we know the exact solution, the correct value of y at $x = 2$ is -0.5. We now wish to see that the error at $t = 2$ divided by h^2 is nearly constant.

```
>   for k from 1 to 8 do
    s:=evalf(2^(-k)):
    sol:=dsolve({diffeq,inits},y(t),numeric,method=classical[heunform],stepsize=s):
    error:=-0.5-subs(sol(2),y(t)):
    ss:=convert(s,string);
    error/s^2;
    r:=convert(",string);
    print('stepsize='.ss,' ratio='.r);
    od:
```

$$stepsize = .5000000000, \quad ratio = .3992994280e - 1$$

$$stepsize = .2500000000, \quad ratio = .4207846720e - 1$$

$$stepsize = .1250000000, \quad ratio = .4054353920e - 1$$

$$stepsize = .6250000000e - 1, \quad ratio = .3939742720e - 1$$

$$stepsize = .3125000000e - 1, \quad ratio = .3876085760e - 1$$

$$stepsize = .1562500000e - 1, \quad ratio = .3842990080e - 1$$

$$stepsize = .7812500000e - 2, \quad ratio = .3826155520e - 1$$

$$stepsize = .3906250000e - 2, \quad ratio = .3817472001e - 1$$

Note that the ratio seems to be approximately constant at 0.038.

8.2   The Taylor Series Method.

One of the disadvantages of the Runge-Kutta method, the "Simpson's rule" of numerical solutions, is that its derivation is somewhat complicated. In this section we present the Taylor series method, which can be used to produce a numerical approximation of any order and which has an easily understandable derivation. In particular, we show how to derive a fourth order Taylor series method using Maple.

We begin by showing how to take higher derivatives of solutions of a first order differential equation

$$y' = \frac{y^2}{2} - \frac{y}{2x}.$$

For example, to take the second derivative of a solution, we differentiate both sides of the differential equation. By the chain rule, this result contains y'. Since the original differential equation tells us the value of y', we can substitute this value for y'. (The Maple command expand will multiply out a product.)

```
>  eqn:=diff(y(x),x)=(y(x)^2-y(x)/x)/2;
```

$$eqn := \frac{\partial}{\partial x} y(x) = \frac{1}{2} y(x)^2 - \frac{1}{2} \frac{y(x)}{x}$$

```
>  diff(rhs(eqn),x);subs(eqn,");
```

$$y(x) \left(\frac{\partial}{\partial x} y(x) \right) - \frac{1}{2} \frac{\frac{\partial}{\partial x} y(x)}{x} + \frac{1}{2} \frac{y(x)}{x^2}$$

$$y(x) \left(\frac{1}{2} y(x)^2 - \frac{1}{2} \frac{y(x)}{x} \right) - \frac{1}{2} \frac{\frac{1}{2} y(x)^2 - \frac{1}{2} \frac{y(x)}{x}}{x} + \frac{1}{2} \frac{y(x)}{x^2}$$

```
>  expand(");  #Second derivative
```

$$\frac{1}{2} y(x)^3 - \frac{3}{4} \frac{y(x)^2}{x} + \frac{3}{4} \frac{y(x)}{x^2}$$

By repeating this bootstrap procedure, we can compute successive derivatives. The successive derivatives are turned into Maple functions $f1, f2, f3, f4$ by means of the unapply(,x,y) command.

```
>  n:=4;
   eq.1:=diff(y(x),x)=(y(x)^2-y(x)/x)/2:
   subs(y(x)=y,rhs(")):
   f.1:=unapply(",x,y):
   eq.1;
   for i from 1 to (n-1) do
   diff(eq.i,x):
   lhs(")=subs(eq.1,rhs(diff(eq.i,x))):
   eq.(i+1):=expand("):
   subs(y(x)=y,rhs(")):
   f.(i+1):=unapply(",x,y):
   print(eq.(i+1));
   od:
```

$$n := 4$$

$$\frac{\partial}{\partial x}\,\mathrm{y}(x) = \frac{1}{2}\,\mathrm{y}(x)^2 - \frac{1}{2}\,\frac{\mathrm{y}(x)}{x}$$

$$\frac{\partial^2}{\partial x^2}\,\mathrm{y}(x) = \frac{1}{2}\,\mathrm{y}(x)^3 - \frac{3}{4}\,\frac{\mathrm{y}(x)^2}{x} + \frac{3}{4}\,\frac{\mathrm{y}(x)}{x^2}$$

$$\frac{\partial^3}{\partial x^3}\,\mathrm{y}(x) = \frac{3}{4}\,\mathrm{y}(x)^4 - \frac{3}{2}\,\frac{\mathrm{y}(x)^3}{x} + \frac{15}{8}\,\frac{\mathrm{y}(x)^2}{x^2} - \frac{15}{8}\,\frac{\mathrm{y}(x)}{x^3}$$

$$\frac{\partial^4}{\partial x^4}\,\mathrm{y}(x) = \frac{3}{2}\,\mathrm{y}(x)^5 - \frac{15}{4}\,\frac{\mathrm{y}(x)^4}{x} + \frac{45}{8}\,\frac{\mathrm{y}(x)^3}{x^2} - \frac{105}{16}\,\frac{\mathrm{y}(x)^2}{x^3} + \frac{105}{16}\,\frac{\mathrm{y}(x)}{x^4}$$

Using the functions for the various derivatives, the general Taylor series of order n for a solution to the differential equation is given as a procedure. (Note the use of the add command below. This is new in Release 4. Also note that when a procedure ends in a semicolon, Maple will parrot it back if the syntax is correct. Once you know that the syntax is correct, you can then change the semicolon to a colon to suppress the additional output, if you wish.)

```
>   T:=proc(x,y,h,k)
    local i;
    y+add(h^i*f.i(x,y)/i!,i=1..k);
    end;
```

$$T := \mathbf{proc}(x,\, y,\, h,\, k)\,\mathbf{local}\,i;\; y + \mathrm{add}(h^i \times f.i(x,\, y)/i!,\, i = 1..k)\,\mathbf{end}$$

Thus, letting h be the step size, the fourth order Taylor series at the point (x, y) for a solution curve, viewed as a function of h is

```
>   T(x,y,h,4);
```

$$y + h\left(\frac{1}{2}\,y^2 - \frac{1}{2}\,\frac{y}{x}\right) + \frac{1}{2}\,h^2\left(\frac{1}{2}\,y^3 - \frac{3}{4}\,\frac{y^2}{x} + \frac{3}{4}\,\frac{y}{x^2}\right) + \frac{1}{6}\,h^3\left(\frac{3}{4}\,y^4 - \frac{3}{2}\,\frac{y^3}{x} + \frac{15}{8}\,\frac{y^2}{x^2} - \frac{15}{8}\,\frac{y}{x^3}\right)$$
$$+ \frac{1}{24}\,h^4\left(\frac{3}{2}\,y^5 - \frac{15}{4}\,\frac{y^4}{x} + \frac{45}{8}\,\frac{y^3}{x^2} - \frac{105}{16}\,\frac{y^2}{x^3} + \frac{105}{16}\,\frac{y}{x^4}\right)$$

Hence, for the fourth order approximation to the solution, we approximate the new value of y in terms of the old values of x and y by using a truncated Taylor series.

```
>   T4:=proc(x)
    local x1,y1,h1:
    x1:=1:y1:=-1:h1:=0.1:
    while x1+h1<=x do
    y1:=T(x1,y1,h1,4):
    x1:=x1+h1;
    od;
    y1;
    end:
```

```
>   T4(2);
```

$$-.5000114875$$

8.3   Finite Difference Method for Boundary Value Problem.

The numerical methods considered thus far have all been applied to initial value problems. Another important class of problems is the *two-point boundary value problems.* In this case we consider a typical example.[1] The following problem, which models the deflection of a horizontal beam subject to uniform (vertical loading) loading intensity w and a horizontal axial force P at the ends:

$$u'' + \frac{P}{EI}u = \frac{wx}{2EI}(L - x), \quad u(0) = 0, \quad u(L) = 0 \tag{8.2}$$

In (8.2), $u = u(x)$ represents the vertical deflection of the beam a distance x from one end of the beam, L is the length of the beam, E is the modulus of elasticity, and I is the moment of inertia. It happens that in this simple case, an exact solution is easily obtained. We do this with Maple using typical values for the parameters, where lengths are in inches and force is in pounds:

```
>   restart:EI:=10^9:L:=300:w:=8:P:=10^5:
```

```
>   diffeq:=(D@@2)(u)(x)+P*u(x)/EI=w*x*(L-x)/(2*EI);
```

$$\textit{diffeq} := (D^{(2)})(u)(x) + \frac{1}{10000}\,u(x) = \frac{1}{250000000}\,x\,(300 - x)$$

```
>   sol:=dsolve({diffeq,u(0)=0,u(L)=0},u(x));
```

$$sol := u(x) = \frac{4}{5} + \frac{3}{250}\,x - \frac{1}{25000}\,x^2 - \frac{4}{5}\cos(\frac{1}{100}\,x) + \frac{4}{5}\,\frac{(-1+\cos(3))\sin(\frac{1}{100}\,x)}{\sin(3)}$$

```
>   uexact:=rhs(sol):plot(uexact,x=0..300);
```

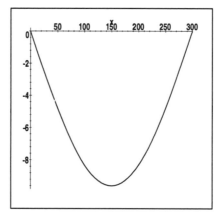

[1]The authors wish to thank Professor Harry Jones of the Texas A&M University Department of Civil Engineering for suggesting this section and the illustrative exercise.

Now it can easily happen that the beam may not have a uniform cross section, or the loading may be a function of position, x. Complications like these lead to problems which are not easily solved analytically. For this reason, approximation methods are needed. We next illustrate an important and widely used approximation technique, the finite difference method. Rather than starting at one end and using a given approximation to the derivative to step our way across, as we have done in the previous differential equations, we form a large system of linear equations, which is solved simultaneously.

We first choose a positive integer n, then divide $[0, L]$ into $n + 1$ subintervals, each of length $h = \dfrac{L}{n+1}$, with endpoints at the points $x_i = ih, i = 0, 1, ..., n + 1$. We next assume that u in (8.2) has four continuous derivatives on $[0, L]$ and replace u'' in (8.2) by

$$u''(x_i) = \frac{u(x_{i-1}) - 2u(x_i) + u(x_{i+1})}{h^2} + O(h^2) \tag{8.3}$$

where we can show by Taylor expansions of $u(x_i + h)$ and $u(x_i - h)$ about x_i that $O(h^2) = -\dfrac{h^2}{12} u^{(4)}(\xi_i)$, for some $\xi \in (x_{i-1}, x_{i+1}), i = 1, 2, ..., n$. This suggests the following approximation to (8.2):

$$\frac{u_{i-1} - 2u_i + u_{i+1}}{h^2} + \frac{P}{EI} u_i = \frac{wx_i(L - x_i)}{2EI}, \quad i = 1, 2, ..., n. \tag{8.4}$$

The equations (8.4) form a linear system of algebraic equations of the form

$$\mathbf{Au} = \mathbf{b}, \tag{8.5}$$

where $\mathbf{A}$ is a tridiagonal $n \times n$ matrix. Such a system is readily solved (see 11.2), for example by Gaussian elimination. This is only the first of many situations in which the techniques of linear algebra play a role in the solution of differential equations.

8.4   Numerical Solution of Systems of First Order Equations

It turns out that systems of equations can be handled by the same numerical methods that are used for scalar equations, with only straightforward modifications. Consider the general first order system of initial value problems written in vector form as

$$\frac{d\mathbf{y}}{dt} = \mathbf{f}(t, \mathbf{y}), \quad \mathbf{y}(t_0) = \mathbf{y0} \tag{8.6}$$

where

$$\mathbf{y} = [y_1, y_2, ..., y_m]^T,$$

$$\mathbf{f}(t, \mathbf{y}) = [f_1(t, \mathbf{y}), f_2(t, \mathbf{y}), ..., f_m(t, \mathbf{y})]^T,$$

t_0 is a fixed real number, and $\mathbf{y0}$ is a given m-vector. (The T superscript denotes "transpose;" i.e., $\mathbf{y}$ is a column vector.)

We begin with Euler's method. Let h be a small positive number, referred to as the *step size*. We define $\mathbf{y}_0 = \mathbf{y0}$ and $t_k = t_0 + hk$. In addition, we write $\mathbf{y}_k$ as an approximation of $\mathbf{y}(t_0 + kh)$. That is,

$$\mathbf{y}_k \approx \mathbf{y}(t_0 + kh).$$

Then we have

$$\frac{\mathbf{y}_{k+1} - \mathbf{y}_k}{h} \approx \mathbf{y}'(t_k) = \mathbf{f}(t_k, \mathbf{y}(t_k))$$

as a motivation for Euler's method for (8.6). Similar to the case for a single equation, Euler's method has the form

$$\mathbf{y}_{k+1} = \mathbf{y}_k + h\mathbf{f}(t, \mathbf{y}_k), \quad k = 0, ..., n - 1 \tag{8.7}$$

While Euler's method is not a very accurate tool for approximating the solution to (8.6), it serves as a simple starting point in understanding numerical approximation of solutions; and it is an easy method to implement.

The following Maple procedure will allow us to perform Euler's method on a system of the form (8.6). Notice the use of the command `evalm` which allows us to work with our variables in vector form.

```
>   restart:
    eul:=proc(f,t0,y0,h,n)
    local i, ti, y;
    y:=array(0..n);
    y[0]:=y0;
    for i from 0 to n-1 do
    ti:=t0+i*h;
    y[i+1]:=evalm(y[i]+h*f(ti,y[i]));
    od;
    y;
    end:
```

Consider the problem of trying to approximate the solution to

$$y_1'(t) = y_2(t) - y_1(t) + e^{-t}, \quad y_2'(t) = -y_1(t) - y_2(t)$$

subject to the initial conditions

$$y_1(0) = -2, \quad y_2(0) = 1$$

This is (8.6) with $t_0 = 0$, $\mathbf{y0} = [-2, 1]^T$ and $\mathbf{f}(t, \mathbf{y}) = [y_2 - y_1 + e^{-t}, -y_1 - y_2]^T$. We have purposely chosen an example for which we can obtain the exact solution, so that we can compare our numerical approximation from Euler's method with the exact solution. To obtain our numerical approximation, we need to write a procedure for the vector field $\mathbf{f}(t, \mathbf{y})$.

```
>   f:=proc(t,y) [y[2]-y[1]+exp(-t),-y[1]-y[2]]end;
```

$$f := \mathbf{proc}(t, y) \, [y_2 - y_1 + \exp(-t), -y_1 - y_2] \, \mathbf{end}$$

Now we are ready to obtain our solution. Let's use a step size of $h = 0.01$, and perform enough iterations of Euler's method to approximate our solution up to time $t = 1$, say 100 iterations. We simply call the procedure `eul` and pass the arguments that it requires. Notice the use of floating point numbers below. **Floating point numbers should always be used in our numerical approximations to avoid Maple's use of symbolic computation.**

```
>   soln:=eul(f,0,[-2.,1.],0.01,100);
```

$$soln := y$$

At first glance it appears as though we have not obtained anything, but this is not the case. All of our approximations are contained in the list soln. For example, if we want the approximation $y_{50} \approx y(0 + (50)0.01) = y(0.5)$, we ask for the value of soln[50].

> evalm(soln[50]);

$$[-.4743433762, 1.043533662]$$

Similarly, if we want the approximation $y_{100} \approx y(0 + (100)0.01) = y(1)$, then we ask for the value of soln[100].

> evalm(soln[100]);

$$[.2326766149, .6481159493]$$

Now let's see how good our approximation is. We can find the exact solution of the system above using dsolve.

> diffeq1:=diff(y1(t),t)=y2(t)-y1(t)+exp(-t);

$$diffeq1 := \frac{\partial}{\partial t} \text{y1}(t) = \text{y2}(t) - \text{y1}(t) + e^{(-t)}$$

> diffeq2:=diff(y2(t),t)=-y1(t)-y2(t);

$$diffeq2 := \frac{\partial}{\partial t} \text{y2}(t) = -\text{y1}(t) - \text{y2}(t)$$

> inits:=y1(0)=-2,y2(0)=1:

> exactsol:=dsolve({diffeq1,diffeq2,inits},{y1(t),y2(t)});

$$exactsol :=$$
$$\{\text{y2}(t) = 2\,e^{(-t)}\cos(t) + 2\,e^{(-t)}\sin(t) - e^{(-t)}, \ \text{y1}(t) = 2\,e^{(-t)}\sin(t) - 2\,e^{(-t)}\cos(t)\}$$

Next, we create Maple functions for each component to use to compare the exact solution with the approximation.

> subs(exactsol,y1(t)):ny1:=unapply(",t);

$$ny1 := t \rightarrow 2\,e^{(-t)}\sin(t) - 2\,e^{(-t)}\cos(t)$$

> subs(exactsol,y2(t)):ny2:=unapply(",t);

$$ny2 := t \rightarrow 2\,e^{(-t)}\cos(t) + 2\,e^{(-t)}\sin(t) - e^{(-t)}$$

We compare our numerical approximation to the true solution in a loop as follows.

> for i from 10 by 10 to 100 do
  print(evalm(soln[i]),[evalf(ny1(i/100)),evalf(ny2(i/100))]);
  od;

$$[-1.617962915, 1.078560571], \ [-1.619967978, 1.076462604]$$

$$[-1.275557755, 1.114821274], \ [-1.279507913, 1.111403924]$$

$$[-.9718634972, 1.116580718], \ [-.9776078486, 1.112496642]$$

$$[-.7054181097, 1.090779002], [-.7127414538, 1.086561092]$$

$$[-.4743433762, 1.043533662], [-.4829888836, 1.039603376]$$

$$[-.2764546928, .9801834591], [-.2861428590, .9768606609]$$

$$[-.1093565648, .9053378417], [-.1198007080, .9028515482]$$

$$[.02947535470, .8229302155], [.0185567304, .8214297819]$$

$$[.1426247449, .7362734052], [.1314984036, .7358397573]$$

$$[.2326766149, .6481159493], [.2215875306, .6487725308]$$

As you can see, the agreement is not great, but it could be a lot worse. Notice that the approximation gets worse as t increases towards 1. This is typical of any numerical method. The approximation is best near the initial t value, and becomes steadily worse as t moves away from this value.

We can execute the same Euler's method solution more easily by appealing to dsolve/numeric.

```
>   meth:=method=classical[foreuler],stepsize=0.01:
```

```
>   sol:=dsolve({diffeq1,diffeq2,inits},{y1(t),y2(t)},numeric,meth);
```

$$sol := \mathbf{proc}(x\_classical) \dots \mathbf{end}$$

We can then find the values of the y's at $t = -.5$ and $t = 1.0$.

```
>   sol(0.5);sol(1.0);
```

$$[t = .5, y2(t) = 1.043533663355845, y1(t) = -.4743433732142408]$$

$$[t = 1.0, y2(t) = .6481159492082968, y1(t) = .2326766166241533]$$

If we want a more accurate numerical solution to a system, we can use the Runge-Kutta-Fehlberg default for dsolve/numeric. This method adapts its step size based on accuracy considerations built into the procedure.

```
>   sol1:=dsolve({diffeq1,diffeq2,inits},{y1(t),y2(t)},numeric);
```

$$sol1 := \mathbf{proc}(rkf45\_x) \dots \mathbf{end}$$

```
>   sol1(0.5);sol1(1.0);
```

$$[t = .5, y2(t) = 1.039603375406419, y1(t) = -.4829888877936588]$$

$$[t = 1.0, y2(t) = .6487725310853097, y1(t) = .2215875256347440]$$

8.5 Two-Dimensional Motion of a Projectile with Drag

In this section, we illustrate how Maple can be used to solve a typical physical problem. Consider a point mass moving through the air under the influence of gravity and a drag force due to air resistance. We take the drag force to be proportional to the square of the speed and acting in the direction opposite to the velocity. Applying Newton's second law, the equation of motion in vector form is:

$$m\frac{d\mathbf{v}}{dt} = -k\mathbf{v}||\mathbf{v}|| - m\mathbf{g} \tag{8.8}$$

Actually, this first order system for the velocity vector $\mathbf{v}(t)$ is a second order system for the position vector $[x(t), y(t)]^T$. In fact, if we write $\mathbf{v}(t) = [vx(t), vy(t)]^T$, then we must have $x'(t) = vx(t)$ and $y'(t) = vy(t)$. In addition, $||\mathbf{v}(t)|| = \sqrt{vx(t)^2 + vy(t)^2}$. Consequently, the equation above (8.8) gives rise to the following first order system in the four unknowns $x(t)$, $y(t)$, $vx(t)$, $vy(t)$:

$$x'(t) = vx(t), \quad y'(t) = vy(t), \tag{8.9}$$

$$vx'(t) = -\frac{k}{m}vx(t)\sqrt{vx(t)^2 + vy(t)^2}, \quad vy'(t) = -\frac{k}{m}vx(t)\sqrt{vx(t)^2 + vy(t)^2} - g \tag{8.10}$$

Let's obtain a numerical solution to this problem. First, assume the point mass is at the origin at time $t = 0$ and is moving with speed $v0$ at an angle α to the x-axis. For simplicity, take $v0 = 100$ ft/sec, $g = 32.2$ ft/sec^2, $\alpha = \pi/4$ and $k/m = 0.0025$. Then the following Maple commands illustrate how to obtain a numerical solution to the system of four first order equations (8.9), (8.10) resulting from equation (8.8), subject to the initial conditions and parameter values given above.

```
>  km:=0.0025:g:=32.2:v0:=100.:alpha:=evalf(Pi/4.):

>  diffeq1:=diff(x(t),t)=vx(t):

>  diffeq2:=diff(y(t),t)=vy(t):

>  diffeq3:=diff(vx(t),t)=-km*vx(t)*sqrt(vx(t)^2+vy(t)^2):

>  diffeq4:=diff(vy(t),t)=-km*vy(t)*sqrt(vx(t)^2+vy(t)^2)-g:

>  sys:=diffeq4,diffeq3,diffeq2,diffeq1:

>  inits:=x(0)=0,y(0)=0,vx(0)=v0*cos(alpha),vy(0)=v0*sin(alpha);
```

$$\mathit{inits} := x(0) = 0, \; y(0) = 0, \; vx(0) = 70.71067811, \; vy(0) = 70.71067813$$

```
>  eqns:={sys,inits}:fcns:={x(t),y(t),vx(t),vy(t)}:

>  mylist:=p*0.2$p=0..5;
```

$$\mathit{mylist} := 0, \; .2, \; .4, \; .6, \; .8, \; 1.0$$

```
> sol:=dsolve(eqns,fcns,numeric,value=array([mylist]));
```

$$
sol := \begin{bmatrix} \begin{bmatrix} & [t,\ \mathrm{vy}(t),\ \mathrm{vx}(t),\ \mathrm{y}(t),\ \mathrm{x}(t)] & & \\ 0 & 70.71067813 & 70.71067811 & 0 & 0 \\ .2 & 61.12544952 & 67.41661011 & 13.17105212 & 13.80493430 \\ .4 & 52.22128686 & 64.54898287 & 24.49551116 & 26.99503393 \\ .6 & 43.87230453 & 62.03564503 & 34.09652356 & 39.64811808 \\ .8 & 35.97781150 & 59.81749852 & 42.07468908 & 51.82895054 \\ 1.0 & 28.45675669 & 57.84492375 & 48.51250909 & 63.59147384 \end{bmatrix} \end{bmatrix}
$$

We next illustrate how to use the numerical solution to obtain a plot of the projectile's trajectory over the first 4 seconds.

```
> sol1:=dsolve(eqns,fcns,numeric);
```

$$
sol1 := \mathbf{proc}(rkf45\_x)\ \dots\ \mathbf{end}
$$

```
> with(plots):odeplot(sol1,[x(t),y(t)],0..4);
```

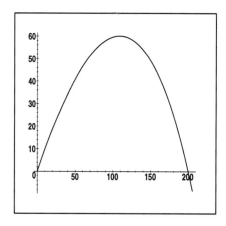

We might want to compare this trajectory with one which ignores air resistance.

```
> ndiffeq3:=diff(vx(t),t)=0;
```

$$
ndiffeq3 := \frac{\partial}{\partial t}\,\mathrm{vx}(t) = 0
$$

```
> ndiffeq4:=diff(vy(t),t)=-g;
```

$$
ndiffeq4 := \frac{\partial}{\partial t}\,\mathrm{vy}(t) = -32.2
$$

```
> nsys:=ndiffeq3,ndiffeq4,diffeq1,diffeq2;
```

$$nsys := \frac{\partial}{\partial t}\,\text{vx}(t) = 0,\ \frac{\partial}{\partial t}\,\text{vy}(t) = -32.2,\ \frac{\partial}{\partial t}\,\text{x}(t) = \text{vx}(t),\ \frac{\partial}{\partial t}\,\text{y}(t) = \text{vy}(t)$$

```
> sol2:=dsolve({nsys,inits},fcns,numeric);
```

$$sol2 := \mathbf{proc}(rkf45\_x) \ldots \mathbf{end}$$

```
> a:=odeplot(sol1,[x(t),y(t)],0..4):
> b:=odeplot(sol2,[x(t),y(t)],0..4):
> with(plots):display([a,b]);
```

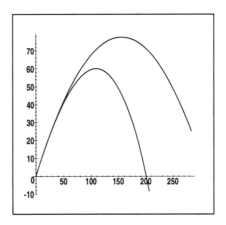

8.6 Exercises

1. In the modified Euler's method developed in Section 1.3, the first y term was calculated with Euler's method and step size h. Would we have been better off to use a higher power of h?

2. Set $h = 0.05$ in each of the three algorithms constructed in Section 1.3, and determine yet another set of approximations to the initial value problem (1.1) that we have been solving numerically.

3. Consider the initial value problem $y' = (t^2 - y^2)\sin y$, $y(0) = -1$.

   (a) Use the method=rkf45 option to solve the equation and find the value $y(2)$.

   (b) Look up dsolve/numeric in the "Help" facility and find out about the method dverk78. Use the method=dverk78 option to solve the equation and find the value $y(2)$.

4. Two other higher order numerical schemes for the initial value problem $y' = f(t,y)$, $y(t_0) = y_0$ are given by

   - $y_{n+1} = y_n + hf\left[t_n + \dfrac{1}{2}h, y_n + \dfrac{1}{2}hf(t_n, y_n)\right]$

- $y_{n+1} = y_n + \dfrac{y'_n + f(t_n + h, y + hy'_n)}{2} h$

For each of these algorithms, write a procedure to calculate the value of y at 0.1, 0.2, 0.3 and 0.4 for $y' = 5t - 3\sqrt{3}, \quad y(0) = 2$.

5. Three procedures defined in this chapter use the argument *stepsize*. It may happen that the distance from *firstx* to *lastx* is *not* an integer multiple of *stepsize*; for example, *firstx* $= 0.0$, *lastx* $= 0.25$, *stepsize* $= 0.1$. What does Maple's output approximate in this case? What does the Release 4 command dsolve/numeric do in this situation, provided that you have specified a classical method such as foreuler (our Euler method) and a given step size?

6. Maple has a time() command which outputs the total CPU time used since the start of a session. Thus, to determine how much CPU time a process takes, invoke this command just before and just after running the process in question. Use this feature of Maple to see how much CPU time each of the three numerical methods take to solve the initial value problem (1.1) one hundred times. Then do the same with the numeric option of dsolve. What happens when the step size is halved?

7. Maple has a taylor command which computes the Taylor series of a function to a prescribed order. With the convert command, Maple can extract the corresponding Taylor polynomial. Use this feature of Maple to compute a third order approximation of $[y(x + h) - y(x)]/h$. Notice this differs from the true derivative of y at the point x by a term of order h. Use this idea to see how good an approximation the symmetric difference quotient (8.1) is to the actual derivative of y.

8. A first order differential equation is of the form $\dfrac{dy}{dx} = f(x, y(x))$. Sometimes a reasonable approximation to the solution of an initial value problem can be constructed by using this equation to calculate the values of higher order derivatives of the unknown solution at the initial x value, and then constructing a Taylor polynomial approximation to the solution. For the intial value problem (1.1), use this method to calculate the sixth order Taylor polynomial approximation to the solution about the point $x = 0$. Compare the value of this Taylor polynomial at $x = 10$ to the other approximations found.

9. Use the *MyModEuler* procedure in Section 1.3 to solve the initial value problem

$$y' = -2ty + \sin t, \ y(0) = 1.$$

Plot your solution on the range t=0..4.0. What interesting event happens?

10. Consider the initial value problem

$$\frac{dy}{dt} = y^3 - 4.56y^2 + 6.9291y - 3.508596, \quad y(0) = 1.48$$

(a) Examine the asymptotic behavior of the solution as t gets large. Discuss the accuracy of the techniques that you use to find your answer.

(b) Suppose that the measurement of the initial condition was only approximate, i.e., maybe $y(0)$ is *not* exactly 1.48. How far off on either side would you have to be to get a different result than the asymptote which you found above?

11. Consider the Airy equation $y'' = xy$ with $y(0) = 0$ and $y'(0) = 1$.

(a) Set `Order:=49;` and use `dsolve` with `type=series` to produce a series solution.

(b) Use `convert(",polynom);` to convert the series solution to a polynomial labelled *sol1*.

(c) Now use `dsolve` with the `numeric` option to produce a numerical solution, *sol2*.

(d) Display both the plot of the polynomial and the plot of the numerical procedure on the same graph for $-8 < x < 2$ and $-2 < y < 2$.

12. Explore the plotting of the `dsolve/numeric` output for a classical method with preset step size. For example,

```
> diffeq:=diff(y(x),x)=y(x);inits:=y(0)=1;

> meth:=method=classical[foreuler];

> val:=value=array([0,0.5,1.0]);

> myoptions:=numeric,meth,stepsize=0.5:

> sol2:=dsolve({diffeq,inits},y(x),myoptions);
```

Observe that one would expect a polygon for a plot. Do you get one?

```
> with(plots):odeplot(sol2,[x,y(x)],0..1);
```

Release 4 has a command to turn off adaptive plotting and only evaluate the expression at certain selected points. Try this command with `odeplot`.

```
> myoptions:=adaptive=false,sample=[0,0.5,1.0]:

> odeplot(sol2,[x,y(x)],0..1,myoptions);
```

Use the delayed evaluation single quotes to avoid an error message about boolean variables in a numeric subroutine, and compute a sequence of points to be plotted.

```
> seq1:='sol2(p*0.5)'$p=0..2;

> seq2:='subs(sol2(p*0.5),[x,y(x)])'$p=0..2;
```

Now plot the points to get the polygon.

```
> plot([seq2]);
```

Now try an alternative method:

```
> odeplot(sol2,[x,y(x)],0..1,numpoints=2);
```

13. Discover a fast implementation of the Runge-Kutta algorithm in Maple. Type `with(share);`, then `readshare(ODE,plots);`. Read about `rungekuttahf` and compare its speed to the implementation in this manual. How many entries are there in `share,index`?

Solve each of the four initial value problems below in three ways: use (`classical[foreuler]`), i.e., Euler's method, improved Euler (`classical[heunform]`), and Runge-Kutta (`classical[rk4]`). Compare the results with the exact solutions (obtained by different methods).

14. Using step sizes $h = 0.1$ and $h = 0.01$, solve

$$x' = 3x - y, \quad x(0) = 0,$$
$$y' = x - 2y, \quad y(0) = 2,$$
$$0 \le t \le 2.$$

15. $\mathbf{x}' = \begin{bmatrix} -2 & 1 \\ -5 & 4 \end{bmatrix} \mathbf{x}, \quad \mathbf{x}(0) = \begin{bmatrix} 1 \\ 3 \end{bmatrix}, \quad 0 \le t \le 1, \quad h = 0.1.$

16. $\mathbf{x}' = \begin{bmatrix} 1 & -5 \\ 1 & -3 \end{bmatrix} \mathbf{x}, \quad \mathbf{x}(0) = \begin{bmatrix} 1 \\ 1 \end{bmatrix}, \quad 0 \le t \le 2, \quad h = 0.05.$

17. $\mathbf{x}' = \begin{bmatrix} 1 & 0 & 0 \\ -4 & 1 & 0 \\ 3 & 6 & 2 \end{bmatrix} \mathbf{x}, \quad \mathbf{x}(0) = \begin{bmatrix} -1 \\ 2 \\ -30 \end{bmatrix}, \quad 0 \le t \le 2, \quad h = 0.02.$

For each of the next two exercises, convert the differential equation to a first order system; find numerical solutions using Euler's method, improved Euler's, and Runge-Kutta; and compare the output from these methods with that obtained using dsolve(,numeric);.

18. $x'' + t^2 x' + 3x = t, \quad x(0) = 1, \quad x'(0) = 2, \quad 0 \le t \le 1.$ Use $h = 0.1$.

19. $y'' + \sin(t + y) = \sin t, \quad y(0) = 1, \quad y'(0) = 0, \quad 0 \le t \le 2.$ Use $h = 0.1$.

20. Using method=classical[rk4] solve the following initial value problem for the Lorentz equations with $h = 0.1$. Plot the solution in the xy-plane and the xz-plane, and also plot x versus t.

$$\frac{dx}{dt} = \sigma(-x + y),$$
$$\frac{dy}{dt} = rx - y - xz,$$
$$\frac{dz}{dt} = -bz + xy,$$
$$x(0) = y(0) = z(0) = 5,$$
$$\sigma = 10, \ b = 8/3, \ r = 28,$$
$$0 \le t \le 20.$$

21. Use the Maple Help facility to learn how adjust the accuracy of dsolve(,numeric).

Chapter 9

Partial Differential Equations

The differential equations considered so far have involved only a single independent variable. In this chapter we turn to problems involving more than one. Examples, all from modeling real physical problems, are:

heat equation Heat conduction in a thin wire.

wave equation Vibration of a flexible string under tension.

Laplace's equation Steady state planar heat distribution with prescribed boundary conditions.

These models illustrate the interplay between theory and applications. Fourier discovered the connection between the problem of heat conduction and trigonometric series, a connection which has benefited both areas. The technique applied to the heat equation turned out also to be useful in the other, very different, areas. Today, Maple is employed to make formerly tedious, sometimes impossibly involved, calculations routine. The ability to work with partial differential equations is new to Release 4.

9.1   The Heat Equation

Suppose a thin insulated wire is placed along the x-axis with $x = 0$ at the left end of the wire and $x = L$ at the right end. Let $u(x, t)$ denote the temperature of the wire at position x, time t. The temperature u satisfies the *heat equation*

$$\frac{\partial u}{\partial t}(x,t) = \beta \frac{\partial^2 u}{\partial x^2}(x,t), \quad 0 < x < L, \quad t > 0.$$

The parameter β, called the *diffusivity*, has units cm^2/s. Typical values at ordinary temperatures are $\beta = 1.1$ for copper and $\beta = 0.0036$ for glass.

We now ask Maple to solve the heat equation.

```
>  restart:
>  pdeheat:=diff(u(x,t),t)-beta*diff(u(x,t),x,x)=0;
```

$$pdeheat := (\frac{\partial}{\partial t}\, \mathrm{u}(x,\, t)) - \beta\,(\frac{\partial^2}{\partial x^2}\, \mathrm{u}(x,\, t)) = 0$$

```
> pdesolve(pdeheat,u(x,t));
```

$$\text{pdesolve}((\frac{\partial}{\partial t}\,u(x,\,t)) - \beta\,(\frac{\partial^2}{\partial x^2}\,u(x,\,t)) = 0,\,u(x,\,t))$$

The fact that Maple returned only the original command indicates that it was unable to solve it. We set the value of the variable infolevel to a high number and then reexecute the command to follow along as Maple explores the various techniques in its repertoire.

```
> infolevel[pdesolve]:=9;
```

$$infolevel_{pdesolve} := 9$$

```
> pdesolve(pdeheat,u(x,t));
```

```
pdesolve/analyze:    equation order    2
pdesolve/analyze:    derivatives in equation:    \{t, x\}
pdesolve/analyze:    not an algebraic equation
pdesolve/analyze:    not an ode in obvious disguise
pdesolve/analyze:    not homogeneous monomial equation
pdesolve/analyze:    linear equation
pdesolve/analyze:    not reducible lin. const. coeff.
pdesolve/analyze:    not lin. const. coeff. 1st order
pdesolve/analyze:    trying algorithms for non-constant coeffs
pdesolve/analyze:    not reducible lin. var. coeff, 1st order
pdesolve/analyze:    not in a 'linear' subset of Lagrange
pdesolve/analyze:    trying to factor (non-commuting) pde
pdesolve/analyze:    cannot factor linear pde in the plane
pdesolve/analyze:    trying algorithms for non-linear
pdesolve/analyze:    not first order equation
pdesolve/analyze:    do not know how to handle equation
pdesolve/exact/2:    Do not know how to handle equation
```

$$\text{pdesolve}((\frac{\partial}{\partial t}\,u(x,\,t)) - \beta\,(\frac{\partial^2}{\partial x^2}\,u(x,\,t)) = 0,\,u(x,\,t))$$

To get debugging tracing information, set infolevel[pdesolve_debug]:=3.

```
> infolevel[pdesolve_debug]:=3;
```

$$infolevel_{pdesolve\_debug} := 3$$

```
> pdesolve(pdeheat,u(x,t));
```

```
 1. exact(diff(u(x,t),t)-beta*diff(diff(u(x,t),x),x) = 0    u(x,t)    )
  2. exact:analyze(DER(u(x,t),[0, 1],[x, t])-beta*DER(u(x,t),[2,
0],[x, t])    [u(x,t)]    \{\}    []    )
pdesolve/analyze:    equation order    2
pdesolve/analyze:    derivatives in equation:    \{t, x\}
pdesolve/analyze:    not an algebraic equation
pdesolve/analyze:    not an ode in obvious disguise
pdesolve/analyze:    not homogeneous monomial equation
pdesolve/analyze:    linear equation
pdesolve/analyze:    not reducible lin. const. coeff.
pdesolve/analyze:    not lin. const. coeff. 1st order
pdesolve/analyze:    trying algorithms for non-constant coeffs
pdesolve/analyze:    not reducible lin. var. coeff, 1st order
pdesolve/analyze:    not in a 'linear' subset of Lagrange
```

```
pdesolve/analyze:    trying to factor (non-commuting) pde
pdesolve/analyze:    cannot factor linear pde in the plane
pdesolve/analyze:    trying algorithms for non-linear
pdesolve/analyze:    not first order equation
pdesolve/analyze:    do not know how to handle equation
  2. exact:analyze --> FAIL
pdesolve/exact/2:    Do not know how to handle equation
  1. exact --> FAIL
```

$$\text{pdesolve}((\frac{\partial}{\partial t}\,u(x,\,t)) - \beta\,(\frac{\partial^2}{\partial x^2}\,u(x,\,t)) = 0,\, u(x,\,t))$$

Since Maple failed to solve the problem on its own, we will now solve it "by hand," letting Maple do some of the calculations. Before we show how to approach the problem, notice that the equation does not completely describe the physical situation. In particular, the *initial* temperature distribution and the *boundary* temperature are not described. One way of adding both to the physical description and to the equation is to describe an initial temperature distribution of $f(x)$, $0 \le x \le L$, and end point temperatures $u(0,t) = u(L,t) = 0$.

$$\frac{\partial u}{\partial t}(x,t) = \beta\frac{\partial^2 u}{\partial x^2}(x,t), \quad 0 < x < L, \quad t > 0,$$
$$u(0,t) = u(L,t) = 0, \quad t \ge 0,$$
$$u(x,0) = f(x), \quad 0 \le x \le L.$$

These equations describe an *initial-boundary value* problem.

Separation of variables

Separation of variables is a technique that is effective in solving several types of PDEs. The idea is to look for solutions of the form $u(x,t) = X(x)T(t)$. Of course, it is unlikely that the solution to the initial-boundary value problem will be of this form; but, since the equation is linear and the boundary conditions are homogeneous, linear combinations of such solutions may be available to solve the more general problem.

Example

Solve the PDE

$$\frac{\partial u}{\partial t}(x,t) = 4\frac{\partial^2 u}{\partial x^2}(x,t), \quad 0 < x < \pi, \quad t > 0,$$
$$u(0,t) = u(\pi,t) = 0, \quad t \ge 0,$$
$$u(x,0) = 2\sin x + \sin 2x, \quad 0 \le x \le \pi.$$

```
>  restart:beta:=4:
>  heat:=diff(u(x,t),t)=beta*diff(u(x,t),x$2);
```
$$heat := \frac{\partial}{\partial t}\,u(x,\,t) = 4\,(\frac{\partial^2}{\partial x^2}\,u(x,\,t))$$

Make the substitution $u(x,t) = X(x)T(t)$.

```
>  subs(u(x,t)=X(x)*T(t),heat):expand(");
```
$$X(x)\,(\frac{\partial}{\partial t}\,T(t)) = 4\,(\frac{\partial^2}{\partial x^2}\,X(x))\,T(t)$$

```
>   "/(4*X(x)*T(t))=K;
```

$$\left(\frac{\frac{\partial}{\partial t} \mathrm{T}(t)}{4\mathrm{T}(t)} = \frac{\frac{\partial^2}{\partial x^2} \mathrm{X}(x)}{\mathrm{X}(x)}\right) = K$$

We can then write this as

$$T'(t) = 4KT(t)$$
$$X''(x) = KX(x)$$

The parameter K must be determined from the ODE

$$X''(x) - KX(x) = 0, \quad X(0) = X(\pi) = 0,$$

a two-point boundary value problem. It is not hard to see that $K < 0$, so we inform Maple with an `assume` command.

```
>   assume(K,negative):
```

```
>   dsolve(diff(X(x),x$2)-K*X(x)=0,X(x));K:='K':
```

$$X(x) = \_C1 \cos(\sqrt{-K^\sim}\,x) + \_C2 \sin(\sqrt{-K^\sim}\,x)$$

Only the sine part is zero at $x = 0$; the zero at π forces $\sqrt{-K}\pi = n\pi$, where n is a positive integer. That is, $K = -n^2$, with the associated $X(x) = \sin nx$. (The special values of K are called *eigenvalues*, and the functions are called *eigenfunctions*.)

Now turn to the T equation: this elementary problem has solution $T(t) = ae^{-4n^2t}$, and the building blocks we hope to use are the functions

$$u_n(x, t) = a_n e^{-4n^2t} \sin nx.$$

It happens that the initial condition $u(x, 0) = 2 \sin x + \sin 2x$ is so simple that one can now solve the problem by inspection. The solution is

$$u(x, t) = 2e^{-4t} \sin x + e^{-16t} \sin 2x.$$

We can now use Maple to verify the solution obtained.

```
>   sol:=u(x,t)=2*exp(-4*t)*sin(x)+exp(-16*t)*sin(2*x);
```

$$sol := \mathrm{u}(x,\, t) = 2\,e^{(-4\,t)} \sin(x) + e^{(-16\,t)} \sin(2\,x)$$

```
>   subs(sol,heat=0):
```

```
>   simplify(");
```

$$(-8\,e^{(-4\,t)} \sin(x) - 16\,e^{(-16\,t)} \sin(2\,x) = -8\,e^{(-4\,t)} \sin(x) - 16\,e^{(-16\,t)} \sin(2\,x)) = 0$$

Thus the solution is verified. The boundary and initial conditions are clearly met.

It is instructive to plot the solution. To determine an appropriate t-interval, note that the term $2e^{-4t}$ (at $x = \pi/2$) is much larger than the other one (with exponent $-16t$). The maximum value for the larger term occurs at $x = \pi/2$, where $\sin x = 1$.

```
>   solve(exp(-4*t)=0.1,t):evalf(");
```

$$.5756462732$$

Thus a t-interval of $[0, 0.6]$ should show practically all values of u which are greater than 0.1, i.e., most of the solution behavior.

```
> see1:=axes=frame,orientation=[-60,75]:
> see2:=style=patchcontour:
> sees:=see1,see2:
> plot3d(rhs(sol),x=0..Pi,t=0..0.6,sees);
```

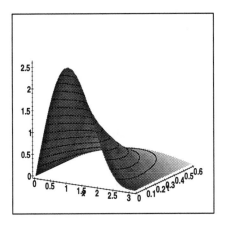

We can also plot the distribution on the boundary.

```
> plot(subs(t=0,rhs(sol)),x=0..Pi);
```

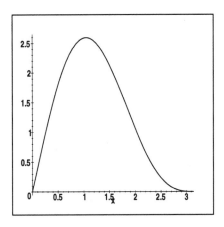

9.2 Fourier Series

Let's return to the original initial-boundary value problem on $[0, L]$. The X-equation is

$$X''(x) - KX(x) = 0, \quad X(0) = X(L) = 0.$$

The eigenvalues are $K_n = -n^2$ with eigenfunctions $\sin(n\pi x/L)$, n a positive integer. It happens that (infinite) *series* of the form

$$\sum_{n=1}^{\infty} b_n \sin\left(\frac{n\pi x}{L}\right)$$

can be used to "represent " almost any function $f(x)$, $0 \le x \le L$. This fact was discovered by the French mathematician Jean Baptiste Joseph Fourier (1768–1830) while studying problems of heat conduction. A formula for the coefficients b_n is

$$b_n = \frac{2}{L} \int_0^L f(x) \sin\left(\frac{n\pi x}{L}\right) dx.$$

Example

Find the Fourier (sine) series of the function f above: $f(x) = 2 \sin x + \sin 2x$ in $[0, \pi]$.

```
> b:=n->(2/Pi)*int(sin(n*x)*(2*sin(x)+sin(2*x)),x=0..Pi);
```

$$b := n \to 2 \frac{\int_0^\pi \sin(n\,x)\,(2\sin(x) + \sin(2\,x))\,dx}{\pi}$$

```
> b(1);
```

$$2$$

```
> b(2);
```

$$1$$

```
> b(3);
```

$$0$$

```
> b(12345);
```

$$0$$

It is not hard to see that $b_n = 0$ for $n \ge 3$. Thus, *the Fourier series of f is f itself.*

Example

Let $f(x) = 4x(x - 1)$ in $0 \le x \le 1$ and $f(x) = (x - 1)(x - \pi)$ if $1 < x \le \pi$. In Maple, this can be coded as:

```
> p:=piecewise(x<1,4*x*(1-x),x<Pi,(x-1)*(Pi-x));
```

$$p := \begin{cases} 4x\,(1 - x) & x < 1 \\ (x - 1)\,(\pi - x) & x < \pi \end{cases}$$

We then turn the Maple expression `piecewise` into a Maple function f and compute a ten term Fourier series approximation, which is then plotted against the original.

```
>   f:=x->p:
```

```
>   b:=n->(2/Pi)*int(f(x)*sin(n*x),x=0..Pi);
```

$$b := n \to 2\,\frac{\displaystyle\int_0^\pi f(x)\sin(n\,x)\,dx}{\pi}$$

```
>   plot({f(x),sum(b(n)*sin(n*x),n=1..10)},x=0..Pi);
```

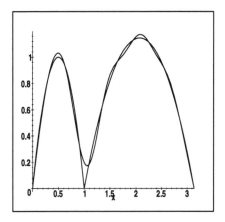

It may be that this approximation is deemed adequate, or that one must increase the number of terms, say to 30:

```
>   plot({f(x),sum(b(n)*sin(n*x),n=1..30)},x=0..Pi);
```

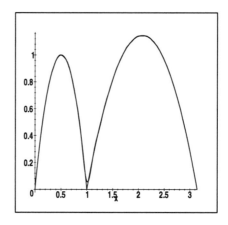

Returning to the heat equation, the solution to the initial value problem is given by

$$u(x,t) = \sum_{n=0}^{\infty} b_n e^{-4n^2 t} \sin nt.$$

Taking infinity ≈ 20, we can repeat the commands above:

```
>   U:=sum(b(n)*exp(-4*n^2*t)*sin(n*x),n=1..20):
>   see1:=axes=frame, orientation=[-20,75]:
>   see2:=style=patchcontour:
>   sees:=see1,see2:
>   plot3d(U,x=0..Pi,t=0..0.6,sees);
```

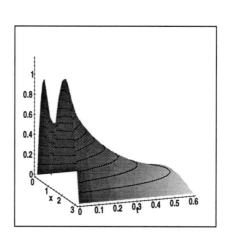

The reader may want to experiment with different viewing angles and plot styles. Also, notice that for all but the very lowest values of t, 20 terms are lavish. We see this by looking at the formula for U. The exponents are $-4n^2t$. Even for $t = .1$, the contribution of other terms to the solution is less than 0.1 for $n > 3$:

```
>   fsolve(exp(-4*n^2*0.1)=0.1,n);
```
$$2.399262956$$

```
>   evalf(b(4)*exp(-4*4^2*0.1));
```
$$.0004712700468$$

It thus appears that $n = 3$ is adequate for $t > .1$:

```
>   U:=sum(b(n)*exp(-4*n^2*t)*sin(n*x),n=1..3):
>   see1:=axes=frame,orientation=[-20,75]:
>   see2:=style=patchcontour:
>   sees:=see1,see2:
>   plot3d(U,x=0..Pi,t=0..0.6,sees);
```

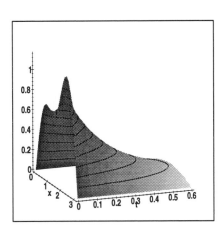

9.2.1 Cosine series

Let us return to the boundary conditions $u(0,t) = u(L,t) = 0$, which originate physically in fixing the temperature at the ends, and replace them by supposing that the wire is insulated at the ends. This leads to the new boundary condition

$$\frac{\partial u}{\partial x}(0,t) = \frac{\partial u}{\partial x}(L,t) = 0.$$

The associated eigenvalue problem, after separating variables with $u(x,t) = X(x)T(t)$, becomes

$$X''(x) - KX(x) = 0, \quad X'(0) = X'(L) = 0,$$
$$T'(t) - \beta KT(t) = 0, \quad t > 0.$$

As before, K must be either zero or negative. In order for the fundamental solution $a \cos \sqrt{-K}t + b \sin \sqrt{-K}t$ of the first differential equation to satisfy the boundary condition, we must have $b = 0$. In other words, the eigenvalues are $-\left(\dfrac{n\pi}{L}\right)^2$, where n is a nonnegative integer. (Zero *is* now allowed.)

Example

Suppose a 3.14 cm copper wire is insulated (including both ends) and has an initial temperature distribution of $u(x,0) = x$. Suppose $\beta = 1.1$ cm^2/sec for copper. Find the temperature $u(x,t)$.

```
>  beta:=1.1:
```

```
>  a:=n->evalf(2/Pi*int(x*cos(n*x),x=0..Pi));
```

$$a := n \rightarrow \mathrm{evalf}\left(2\,\frac{\displaystyle\int_0^{\pi} x\cos(n\,x)\,dx}{\pi}\right)$$

```
>  N:=13:
```

```
>  U:=a(0)/2+sum(a(n)*exp(-1.1*n^2*t)*cos(n*x),n=1..N):
```

```
>  see1:=axes=frame,orientation=[-5,60]:
```

```
>  see2:=style=patchcontour,numpoints=1000:
```

```
>  sees:=see1,see2:
```

```
>  plot3d(U,x=0..Pi,t=0..3,sees);
```

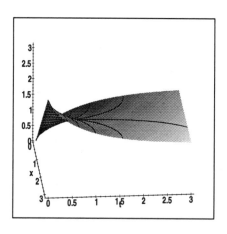

Can you guess what the steady state temperature distribution is? That is, what is $\lim_{t \to \infty} u(x,t)$?

9.2.2 General Fourier Series

The two examples of trigonometric series (the sine series and the cosine series) arise naturally from the initial-boundary value problem following separation of variables. Other boundary conditions can easily be imagined: for example, the temperature could be fixed at one end and the other end could be insulated. In this more general setting, trigonometric series with both sines and cosines are required.

A natural interval on which to study trigonometric series has length 2π, because this is the period of the functions. Thus intervals $[0, 2\pi]$ or $[-\pi, \pi]$ are candidates. Start with a function f which is periodic with period 2π: $f(x + 2\pi) = f(x)$ for all x. Under very mild conditions on the function f, the Fourier series converges to the function at points where the function is continuous; and we have

$$f(x) = \frac{a_0}{2} + \sum_{n=1}^{\infty} a_n \cos nx + b_n \sin nx,$$

where the coefficients are given by the *Euler formulas*[1]

$$a_n = \frac{1}{\pi} \int_0^{2\pi} f(x) \cos nx \, dx, \quad n = 0, 1, \ldots,$$

$$b_n = \frac{1}{\pi} \int_0^{2\pi} f(x) \sin nx \, dx, \quad n = 1, 2, \ldots.$$

Since all that is involved here is integration, Maple would seem to be of great help in the calculations.

Example

Define $f(x) = x$, $0 \le x < 2\pi$, $f(x + 2\pi) = f(x)$, for all x. Find the Fourier series of f.

```
>   a:=n->evalf(1/Pi*int(x*cos(n*x),x=0..2*Pi));
```

$$a := n \rightarrow \text{evalf} \left(\frac{\int_0^{2\pi} x \cos(n\,x)\,dx}{\pi} \right)$$

```
>   b:=n->evalf(1/Pi*int(x*sin(n*x),x=0..2*Pi));
```

$$b := n \rightarrow \text{evalf} \left(\frac{\int_0^{2\pi} x \sin(n\,x)\,dx}{\pi} \right)$$

---

[1] Although the Swiss mathematician Leonhard Euler (1707–1783) knew the formulas, he did not grasp the significance to these applications, nor the incredible richness of the subject.

```
>  N:=6:
>  funplt:=plot(x,x=0..2*Pi,scaling=constrained):
>  fser:=a(0)/2+sum(a(n)*cos(n*x)+b(n)*sin(n*x),n=1..N):
>  serplt:=plot(fser,x=0..2*Pi,scaling=constrained):
>  with(plots):display({funplt,serplt});
```

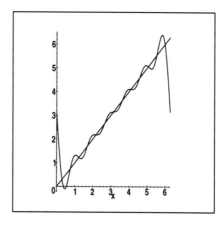

Because the function is discontinuous (at the end points) but the sum is continuous, the "fit" at the end points is poor. Here is an example using a smooth function:

Example

Define $f(x) = x^2(x - 2\pi)^2$, $0 \le x < 2\pi$, $f(x + 2\pi) = f(x)$, for all x. Find the Fourier series of f.

```
>   f:=x^2*(x-2*Pi)^2:
>   a:=n->evalf(1/Pi*int(f*cos(n*x),x=0..2*Pi));
```

$$a := n \to \text{evalf}\left(\frac{\int_0^{2\pi} f\cos(n\,x)\,dx}{\pi}\right)$$

```
>   b:=n->evalf(1/Pi*int(f*sin(n*x),x=0..2*Pi));
```

$$b := n \to \text{evalf}\left(\frac{\int_0^{2\pi} f\sin(n\,x)\,dx}{\pi}\right)$$

```
>  N:=2:
```

```
>  fser:=a(0)/2+sum(a(n)*cos(n*x),n=1..N):
```

```
>  funplt:=plot(f,x=0..2*Pi):
```

```
>  serplt:=plot(fser,x=0..2*Pi):
```

```
>  with(plots):display({funplt,serplt});
```

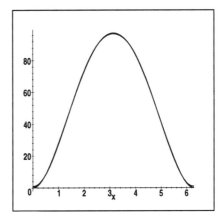

Note the following facts: first, the function is *even*, so the sine integrals all are zero. Next, the approximation is excellent, even for small values of N, as the graph of the difference shows with $N = 2$. The absolute error is largest at the end points.

```
>  N:=10:
```

```
>  fser:=a(0)/2+sum(a(n)*cos(n*x),n=1..N):
```

```
>  plot(f-fser,x=0..2*Pi);
```

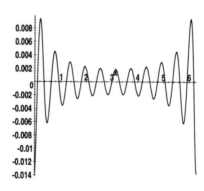

The reader may wish to experiment with other values of N.

Next, we consider the "square wave" function.

Example

Let $f(x) = 1, 0 < x < \pi$, $f(x) = -1, -\pi < x < 0$, $f(0) = f(\pi) = 0$, and $f(x + 2\pi) = f(x)$. In the interval $(-\pi, \pi)$, the function can be written in Maple using `piecewise`:

```
>   piecewise(x<0,-1,x=0,0,x>0,1);f:=unapply(",x):
```

$$\begin{cases} -1 & x < 0 \\ 0 & x = 0 \\ 1 & 0 < x \end{cases}$$

Notice that this function is *odd*, so that the cosine coefficients a_n are zero. Also,

$$b_n = \frac{1}{\pi} \int_{-\pi}^{\pi} f(x) \sin nx \, dx = \frac{2}{\pi} \int_0^{\pi} f(x) \sin nx \, dx.$$

This is exactly the formula for the sine coefficient b_n which we encountered earlier. In Maple:

```
>   b:=n->2/Pi*int(sin(n*x),x=0..Pi);
```

$$b := n \rightarrow 2 \frac{\displaystyle\int_0^{\pi} \sin(n\,x)\,dx}{\pi}$$

```
>   N=10:
```

```
>   sers:=sum(b(n)*sin(n*x),n=1..N):
```

```
>   plot({f(x),sers},x=-Pi..Pi);
```

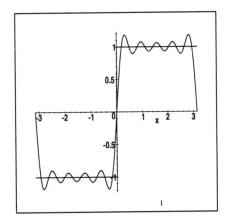

The "overshoot" at the discontinuities of the approximation is known as Gibb's phenomenon after the American mathematical physicist Josiah Gibbs (1839–1903). It amounts to about 8%, and persists as N gets larger:

```
>  N:=30:

>  sers:=sum(b(n)*sin(n*x),n=1..N):

>  plot({f(x),sers},x=-Pi..Pi,numpoints=300);
```

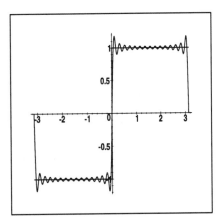

9.3 The Wave Equation

Consider the motion of a horizontal string with density ρ which is fixed at both ends under tension T. Assume the string is perfectly flexible and that it moves only in the vertical direction. Assume that gravity is negligible.

Then the following equation, called the (one dimensional) *wave equation*, holds:

$$\frac{\partial^2 u}{\partial x^2}(x,t) = \frac{1}{\alpha^2}\frac{\partial^2 u}{\partial t^2}(x,t), \quad 0 < x < L, \quad t > 0,$$
$$u(0,t) = u(L,t) = 0, \quad t \geq 0,$$
$$u(x,0) = f(x), \quad 0 \leq x \leq L,$$
$$\frac{\partial u}{\partial t}(x,0) = g(x), \quad 0 < x < L.$$

The constant $\alpha^2 = T/\rho$ is usually very large (which is why gravity can be neglected). Let us see if Maple can be of some assistance:

```
>  restart:

>  pdewave:=diff(u(x,t),x,x)-(1/alpha^2)*diff(u(x,t),t,t)=0;
```

$$pdewave := (\frac{\partial^2}{\partial x^2}u(x,\,t)) - \frac{\frac{\partial^2}{\partial t^2}u(x,\,t)}{\alpha^2} = 0$$

```
>  pdesolve(pdewave,u(x,t));
```

$$u(x,\,t) = \_F1(t\,\alpha + x) + \_F2(t\,\alpha - x)$$

Maple knows the solution, known as *D'Alembert's solution*.[2]   (The symbols $\_F1$ and $\_F2$ are names of arbitrary *functions*, much like the arbitrary *constants* which were encountered in the solution of ordinary differential equations.) Use A and B for these two functions; it is easy to see that $u(x,t) = A(x+\alpha t)+B(x-\alpha t)$ is a solution if A and B are arbitrary twice differentiable functions. D'Alembert's solution is easily derived using the change of variables $w = x + \alpha t$ and $z = x - \alpha t$, which changes the original equation to

$$\frac{\partial^2 v}{\partial z \partial w} = 0, \quad v(w,z) = u(x,t),$$

after applying the chain rule.

The problem now is to determine A and B to satisfy the initial and boundary conditions.
First the initial conditions:

```
>  restart:

>  eq1:=A(x)+B(x)=f(x);
```

$$eq1 := A(x) + B(x) = f(x)$$

```
>  eq2:=alpha*diff(A(x),x)-alpha*diff(B(x),x)=g(x);
```

$$eq2 := \alpha\,(\frac{\partial}{\partial x}A(x)) - \alpha\,(\frac{\partial}{\partial x}B(x)) = g(x)$$

```
>  sol:=dsolve({eq1,eq2},{A(x),B(x)}):

>  ai:=expand(sol[1]);
```

$$ai := A(x) = \_C2 + \frac{1}{2}\frac{\int g(x) + (\frac{\partial}{\partial x}f(x))\,\alpha\,dx}{\alpha}$$

---

[2] After the French mathematician Jean le Rond D'Alembert (1717–1783) .

```
>  bi:=expand(sol[2]);
```

$$bi := \mathrm{B}(x) = \_C1 + \frac{1}{2}\frac{\int (\frac{\partial}{\partial x}\,\mathrm{f}(x))\,\alpha - \mathrm{g}(x)\,dx}{\alpha}$$

The integral of a sum is the sum of the integrals; however, since Maple cannot give a closed form answer to the second integral, the expression returns unevaluated. Hence we proceed by hand. Let

$$G(x) = \frac{1}{\alpha}\int_0^x g(y)\,dy.$$

The above solutions can be written as:

$$A(x) = \tfrac{1}{2}(f(x) + G(x) + C_1), \quad 0 < x < L,$$
$$B(x) = \tfrac{1}{2}(f(x) - G(x) - C_2), \quad 0 < x < L.$$

We now need to determine extensions of functions f and G to all real numbers so that the boundary conditions

$$u(x,0) = A(\alpha t) + B(-\alpha t) = 0,$$
$$u(L,t) = A(L + \alpha t) + B(L - \alpha t) = 0$$

are satisfied. The solution to this problem is to let f be the odd extension of f on the interval $(-L, L]$ and then extend this to be periodic with period $2L$. Also extend G to the interval $[-L, L]$ to be an even function, then extend to all real numbers to have period $2L$. Finally, the solution to the initial-boundary value problem is

$$u(x,t) = \frac{1}{2}(f(x + \alpha t) + f(x - \alpha t) + G(x + \alpha t) - G(x - \alpha t)).$$

Example

A string with $\alpha = 3 \times 10^2$ m/s is 1 m long, and is plucked in the center by displacing the string 0.1 m and releasing. Determine the motion of the string.

The statement implies that $g(x) = 0$, so $G(x) \equiv 0$, too. A formula for $f(x)$ in Maple is

```
>  f:=proc(x)
   if x<0 then -f(-x)
   elif x>=2 then f(x-2)
   elif x>=1 then -f(x-1)
   elif x>=0.5 then 0.2-x/5
   else 0.2*x
   fi;
   end:
```

We can plot this procedure. Note the syntax that is necessary to do so, since we are plotting a Maple procedure, rather than a Maple expression. It is clear from the symmetry that the integral over an entire period is zero.

```
>  plot(f,0..2);
```

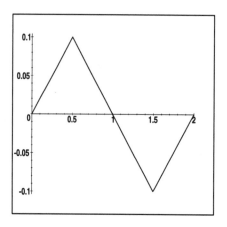

Now we can define $u(x, t)$.

```
>   alpha:=3*10^2:
>   u:=(x,t)->(f(x+alpha*t)+f(x-alpha*t))/2;
```

$$u := (x,\ t) \rightarrow \frac{1}{2}\, \mathrm{f}(x + \alpha\, t) + \frac{1}{2}\, \mathrm{f}(x - \alpha\, t)$$

While running a Maple session, we can click the mouse inside the plot area. The menu will change to the one appropriate for plots. In that menu, one can click on "animate" and see the vibration of the string. (Note that no plot is generated by a call to plots[animate] when one selects Export As LaTeX from the File entry in the menu, even though a plot is shown in the Maple worksheet.)

```
>   plots[animate](u,0..1,0..0.1,frames=200,color=blue);
```

Separation of Variables

In the initial-value problem

$$\frac{\partial^2 u}{\partial x^2}(x, t) = \frac{1}{\alpha^2}\frac{\partial^2 u}{\partial t^2}(x,t), \quad 0 < x < L, \quad t > 0,$$
$$u(0, t) = u(L, t) = 0, \quad t \geq 0,$$

the substitution $u(x, t) = X(x)T(t)$ leads to

$$X''(x)T(t) = \frac{1}{\alpha^2}X(x)T''(t), \quad 0 < x < L, \quad t > 0,$$
$$\frac{X''(x)}{X(x)} = \frac{1}{\alpha^2}\frac{T''(t)}{T(t)} = -\lambda^2,$$
$$X(0) = X(L) = 0.$$

The X-equation leads to the eigenvalues $\lambda_n = \dfrac{n\pi}{L}$ as before. The T-equation is similar, with solution

$$T(t) = a_n \cos \lambda_n \alpha t + b_n \sin \lambda_n \alpha t.$$

The u_n are, accordingly,

$$u_n(x,t) = \sin \lambda_n x \left(a_n \cos \lambda_n \alpha t + b_n \sin \lambda_n \alpha t \right),$$

and we hope to satisfy the initial conditions

$$u(x,0) = f(x), \quad 0 < x < L,$$
$$\frac{\partial u}{\partial t}(x,0) = g(x), \quad 0 < x < L.$$

by selecting coefficients a_n and b_n in the expression

$$u(x,t) = \sum_{n=1}^{\infty} \sin \lambda_n x \left(a_n \cos \lambda_n \alpha t + b_n \sin \lambda_n \alpha t \right).$$

This leads to formulas

$$a_n = \frac{2}{L} \int_0^L f(x) \sin \left(\frac{n\pi x}{L} \right) dx,$$
$$b_n = \frac{2}{n\pi \alpha} \int_0^L g(x) \sin \left(\frac{n\pi x}{L} \right) dx.$$

Example

Return to the earlier wave equation example, and this time use the Fourier series technique. Because $g(x) \equiv 0$ the coefficients b_n are all zero. Recall that $L = 1$ and the initial deflection is $\dfrac{1}{10}$. In Release 4, the `assume` facility is much stronger; and Maple now knows that integral multiples of π have sine equal to 0.

```
>   restart:
>   assume(n,integer):
>   first:=int(x*sin(n*Pi*x),x=0..1/2):
>   second:=int((1-x)*sin(n*Pi*x),x=1/2..1):
>   4/10*(first+second):
>   simplify("):
>   aa:=subs(n='n',"):n:='n':
>   a:=unapply(aa,n);
```

$$a := n \to \frac{2}{5} \frac{\sin(\frac{1}{2} n \pi)}{n^2 \pi^2}$$

```
>  alpha:=300:
```

Notice that if n is even then $a_n = 0$; this somewhat speeds the calculations in the series.

```
>  N:=12:   # Actually get 25 terms
```

```
>  s:=n->a(2*n+1)*sin((2*n+1)*Pi*x)*cos((2*n+1)*Pi*alpha*t);
```

$$s := n \rightarrow a(2n+1) \sin((2n+1)\pi x) \cos((2n+1)\pi \alpha t)$$

```
>  add(s(n),n=0..N):mysum:=":
```

```
>  u:=unapply(mysum,x,t):
```

```
>  plot(u(x,0),x=0..1);
```

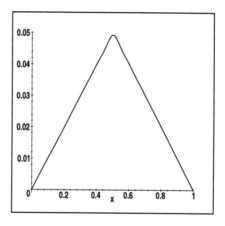

```
>  plots[animate](u(x,t),x=0..1,t=0..1,frames=200);
```

Separation of variables apparently gives the same solution as before (using D'Alembert's solution). The main difference is the time required to perform the Fourier series summations and the fact that the partial sum of the Fourier series is just a trigonometric polynomial, whereas the actual solution is not differentiable everywhere.

9.4 Laplace's Equation

To this point, all equations have been one-dimensional. (Although there are two variables, x and t, there has been only one *spatial* dimension.) The other dimension t represents time, a *temporal* dimension. The change from one to two or three dimensions is not difficult, although the physical interpretation of the parameter in the PDE and the character of the boundary changes dramatically. Symbolically, simply replace $\dfrac{\partial^2 u}{\partial x^2}$ by the

Laplacian of u:

$$\Delta u(x, y) = \frac{\partial^2 u}{\partial x^2} + \frac{\partial^2 u}{\partial y^2} \quad \text{in two dimensions,}$$

$$\Delta u(x, y, z) = \frac{\partial^2 u}{\partial x^2} + \frac{\partial^2 u}{\partial y^2} + \frac{\partial^2 u}{\partial z^2} \quad \text{in three dimensions.}$$

Space does not permit us to study the heat and wave equation in two- and three-dimensions. The biggest change is in the spatial boundary conditions and the choice of a coordinate system, and these are well illustrated by the Laplace[3] equation,[4] $\Delta u = 0$. In one dimension, the Laplace equation is, of course, an ODE: $u''(x) = 0$, with solution $u(x) = c_1 + c_2 x$.

In two dimensions, Laplace's equation is satisfied inside a plane region R with conditions specified on the boundary ∂R of R. Two basic types of boundary conditions are *Dirichlet*[5] boundary conditions, in which the value of u is given on ∂R

$$y(x, y) = f(x, y) \quad \text{on} \quad \partial R,$$

and *Neumann*[6] boundary conditions, in which the directional derivative $\partial u / \partial n$ along the outward normal to the boundary is specified:

$$\frac{\partial u}{\partial n}(x, y) = g(x, y) \quad \text{on} \quad \partial R.$$

Sometimes both types of boundary conditions, i.e., *mixed* boundary conditions are given.

In this section we will use separation of variables in rectangular and circular two dimensional domains to solve various boundary value problems.

A Rectangular Domain

In this example, the domain D is a rectangle with u specified on the top and bottom and the left and right boundaries insulated.

$$\frac{\partial^2 u}{\partial x^2} + \frac{\partial^2 u}{\partial y^2} = 0, \quad 0 < x < a, \quad 0 < y < b,$$

$$\frac{\partial u}{\partial x}(0, y) = \frac{\partial u}{\partial x}(a, y) = 0, \quad 0 \le y \le b,$$

$$u(x, b) = 0, \quad u(x, 0) = f(x), \quad 0 \le x \le a.$$

Separating variables, we let $u(x, y) = X(x)Y(y)$:

$$X''(x)Y(y) + X(x)Y''(y) = 0.$$

As before,

$$\frac{X''(x)}{X(x)} = -\frac{Y''(y)}{Y(y)} = K.$$

This leads to the ODEs

$$X''(x) - KX(x) = 0$$
$$Y''(y) + KY(y) = 0$$

---

[3] After [Marquis de] Pierre Simon Laplace (1749–1827), the French astronomer and mathematician.

[4] Also called the *potential equation*

[5] After the German mathematician Peter Gustav Lejune Dirichlet (1805–1859). Again illustrating the interplay between applications and mathematics, Dirichlet discovered the now accepted definition of a function while considering the problem of convergence of Fourier series.

[6] After the German mathematician Karl Gottfried Neumann (1832–1925).

The first boundary condition gives

$$X'(0) = X'(a) = 0,$$

so the eigenvalue problem must have $K = K_n = -(n\pi/a)^2$, with corresponding eigenfunction

$$X_n(x) = a_n \cos\left(\frac{n\pi x}{a}\right).$$

```
>  yde:=diff(Y(y),y,y)-(n*Pi/a)^2*Y(y)=0;
```

$$yde := (\frac{\partial^2}{\partial y^2} Y(y)) - \frac{n^2 \pi^2 Y(y)}{a^2} = 0$$

```
>  dsolve(yde,Y(y));
```

$$Y(y) = \_C1\, e^{(\frac{n\pi y}{a})} + \_C2\, e^{(-\frac{n\pi y}{a})}$$

```
>  convert(",trig);
```

$$Y(y) = \_C1\left(\cosh(\frac{n\pi y}{a}) + \sinh(\frac{n\pi y}{a})\right) + \_C2\left(\cosh(\frac{n\pi y}{a}) - \sinh(\frac{n\pi y}{a})\right)$$

```
>  collect(",{sinh,cosh});
```

$$Y(y) = (\_C1 - \_C2)\sinh(\frac{n\pi y}{a}) + (\_C1 + \_C2)\cosh(\frac{n\pi y}{a})$$

```
>  subs(_C1+_C2=B,_C1-_C2=A,");
```

$$Y(y) = A\sinh(\frac{n\pi y}{a}) + B\cosh(\frac{n\pi y}{a})$$

This works for $n > 0$. For $n = 0$, the solution is $Y_0(y) = A_0 + B_0 y$. Now look at the $y = b$ boundary condition. Rather than solve the equation as it is written, first note that Y_n can be written as

$$Y_n(y) = C_n \sinh\left(\frac{n\pi}{a}(y + D_n)\right).$$

```
>  ayeq:=rhs(")=C*sinh(y+D);
```

$$ayeq := A\sinh(\frac{n\pi y}{a}) + B\cosh(\frac{n\pi y}{a}) = C\sinh(y + D)$$

```
>  expand(");
```

$$A\sinh(\frac{n\pi y}{a}) + B\cosh(\frac{n\pi y}{a}) = C\sinh(y)\cosh(D) + C\cosh(y)\sinh(D)$$

```
>  meq:={A=C*cosh(D),B=C*sinh(D)};
```

$$meq := \{B = C\sinh(D),\ A = C\cosh(D)\}$$

Dividing the two equations, we get $D = \tanh^{-1}(B/A)$ and $C = B/\cosh D$. (If $A = 0$ the expression is already in the form $C \sinh(y + D)$.) With this representation, we satisfy the $y = b$ boundary condition with

$$Y_0 = B_0(y - b)$$
$$Y_n = C_n \sinh\left(\frac{n\pi}{a}(y - b)\right)$$

This leads to the combined X and Y terms

$$u_0(x, y) = E_0(y - b),$$
$$u_n(x, y) = E_n \cos\left(\frac{n\pi x}{a}\right) \sinh\left(\frac{n\pi}{a}(y - b)\right),$$

where the E_n are constants to be determined.

The remaining boundary condition is simply $u(x, 0) = f(x)$. Notice the series with $y = 0$ is a cosine series. If we expand $f(x)$ in a cosine series

$$f(x) = \frac{a_0}{2} + \sum_{n=1}^{\infty} a_n \cos\left(\frac{n\pi x}{a}\right),$$

then the formula for a_n is

$$a_n = \frac{2}{a} \int_0^a f(x) \cos\left(\frac{n\pi x}{a}\right) dx.$$

From this point it is easy to find E_n.

Example

In the above setting, suppose $a = \pi$, $b = 1$, and $f(x) = 1 - \cos 2x$. (Note that $f(x) = 2\sin^2 x$.) The Fourier cosine series for f is simply f (that is, no integration is necessary).

```
u:=(x,y)->1-y+(-1)/sinh(-2)*cos(2*x)*sinh(2*(y-1))
```

$$u := (x, y) \to 1 - y - \frac{\cos(2x) \sinh(2y - 2)}{\sinh(-2)}$$

```
>  plot(u(x,0),x=0..Pi);
```

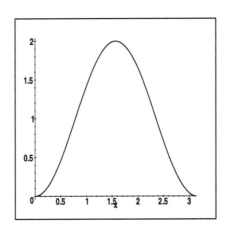

```
>  plot3d(u(x,y),x=0..Pi,y=0..1);
```

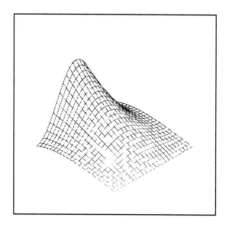

```
>  with(plots):
```

```
>  contourplot(u(x,y),x=0..Pi,y=0..1);
```

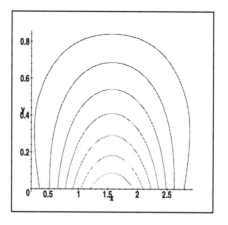

```
>  gradplot(u(x,y),x=0..Pi,y=0..1);
```

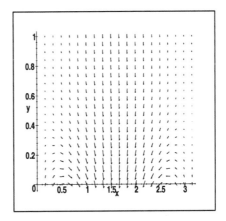

The gradient field is orthogonal to the family of level curves:

```
>  lvl:=contourplot(u(x,y),x=0..Pi/2,y=0..1):
>  dir:=gradplot(u(x,y),x=0..Pi/2,y=0..1):
>  display({lvl,dir},scaling=constrained);
```

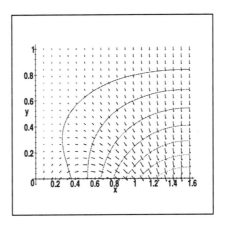

A Circular Domain

Now consider the Dirichlet problem in the unit circle $x^2 + y^2 < 1$. A little experimentation might convince you that separation of variables by the substitution $u(x, y) = X(x)Y(y)$ fails. In a sense, for separation of variables to work, the region has to be rectangle-like. The way out in this case is to change to polar coordinates r, θ. Maple can help:

```
>  restart:
```

```
> Laplacian:=diff(u(x,y),x,x)+diff(u(x,y),y,y)=0;
```

$$Laplacian := (\frac{\partial^2}{\partial x^2} u(x, y)) + (\frac{\partial^2}{\partial y^2} u(x, y)) = 0$$

```
> with(DEtools):
```

We will use PDEchangecoords to change from rectangular to polar coordinates: (Note that one of the restrictions on the Release 4 PDEchangecoords command is that the new coordinate system must be orthogonal.)

```
> PDEchangecoords(Laplacian,[x,y],polar,[r,theta]);
```

$$\frac{(\frac{\partial}{\partial r} u(r, \theta)) r + (\frac{\partial^2}{\partial \theta^2} u(r, \theta)) + (\frac{\partial^2}{\partial r^2} u(r, \theta)) r^2}{r^2} = 0$$

```
> polop:=expand(");
```

$$polop := \frac{\frac{\partial}{\partial r} u(r, \theta)}{r} + \frac{\frac{\partial^2}{\partial \theta^2} u(r, \theta)}{r^2} + (\frac{\partial^2}{\partial r^2} u(r, \theta)) = 0$$

So, in polar coordinates,

$$\Delta u = \frac{\partial^2 u}{\partial r^2} + \frac{1}{r} \frac{\partial u}{\partial r} + \frac{1}{r^2} \frac{\partial^2 u}{\partial \theta^2}.$$

The interior of the unit disk is described mathematically by $r < 1$. In the polar coordinate system, the angle θ is not unique. Let us put a "cut" in the plane along the negative x-axis so that θ is restricted to the interval $-\pi < \theta \leq \pi$. This in not a physically real boundary, and no boundary conditions are associated with $\theta = \pi$. Now try to separate variables: $u(r, \theta) = R(r)\Theta(\theta)$:

```
> subs(u(r,theta)=R(r)*Theta(theta),polop);
```

$$\frac{\frac{\partial}{\partial r} R(r) \Theta(\theta)}{r} + \frac{\frac{\partial^2}{\partial \theta^2} R(r) \Theta(\theta)}{r^2} + (\frac{\partial^2}{\partial r^2} R(r) \Theta(\theta)) = 0$$

```
> simplify(r^2*");
```

$$(\frac{\partial}{\partial r} R(r)) \Theta(\theta) r + R(r) (\frac{\partial^2}{\partial \theta^2} \Theta(\theta)) + (\frac{\partial^2}{\partial r^2} R(r)) \Theta(\theta) r^2 = 0$$

```
> "/(R(r)*Theta(theta)):expand(");
```

$$\frac{(\frac{\partial}{\partial r} R(r)) r}{R(r)} + \frac{\frac{\partial^2}{\partial \theta^2} \Theta(\theta)}{\Theta(\theta)} + \frac{(\frac{\partial^2}{\partial r^2} R(r)) r^2}{R(r)} = 0$$

We have

$$\frac{r^2 R''(r) + r R'(r)}{R(r)} = -\frac{\Theta''(\theta)}{\Theta(\theta)} = \lambda$$

where λ is a number to be determined. The Θ-equation $\Theta'' + \lambda\Theta = 0$ has solution

$$\Theta(\theta) = a \cos \sqrt{\lambda}\theta + b \sin \sqrt{\lambda}\theta.$$

However, we know the solution must be periodic with period 2π, so that $\lambda = n^2$ with n a non-negative integer.
The R equation is now

$$r^2 R''(r) + r R(r) - n^2 R(r) = 2.$$

This turns out to be a *Cauchy-Euler* equation, and Maple can solve it:

```
> CauEul:=r^2*diff(R(r),r,r)+r*diff(R(r),r)-n^2*R(r)=0;
```

$$CauEul := r^2 \left(\frac{\partial^2}{\partial r^2} R(r)\right) + r \left(\frac{\partial}{\partial r} R(r)\right) - n^2 R(r) = 0$$

```
> dsolve(CauEul,R(r));
```

$$R(r) = \_C1\, r^n + \_C2\, r^{(-n)}$$

The r^{-n} term is singular at $r = 0$, so we discard this term and select $R(r) = r^n$. Thus, the solution to the Dirichlet problem in the unit circle is obtained in two steps:

1. Find the general Fourier series of the function $f(\theta)$ on the boundary.

2. Multiply the nth term by r^n.

Couldn't be simpler!

Example

Find the steady-state temperature distribution inside the unit circle when the boundary temperature is held at 1°C on the right half and 0°C on the left half of the circle.

Here we have $f(\theta) = 1$ in $(-\pi/2, \pi/2)$ and 0 otherwise. This is an even function, so the sine terms in the Fourier series are all zero. (Think of the unit disk as being a *rectangle* with $0 \leq r \leq 1$ as the vertical range and $-\pi/2 < \theta < \pi/2$ as the horizontal range. Now the function f looks even.)

```
> restart;
> (1/Pi)*int(cos(n*theta),theta=-Pi/2..Pi/2);
```

$$2\,\frac{\sin(\frac{1}{2}\pi n)}{\pi n}$$

```
> N:=15:
> u:=(r,theta)->(1/2)+sum(r^n*cos(n*theta)*2/(n*Pi)*sin(n*Pi/2),n=1..N);
```

$$u := (r, \theta) \rightarrow \frac{1}{2} + \left(\sum_{n=1}^{N} \left(2\,\frac{r^n \cos(n\,\theta)\sin(\frac{1}{2}n\,\pi)}{n\,\pi}\right)\right)$$

```
> plot3d(u(r,theta),r=0..1,theta=-Pi..Pi);
```

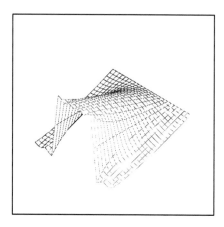

The temperature distribution is supposed to be 1 on the interval $-\pi/2 < \theta < \pi/2$ and 0 elsewhere. Here is a plot of the Fourier approximation.

```
> plot(u(1,t),t=-Pi..Pi);
```

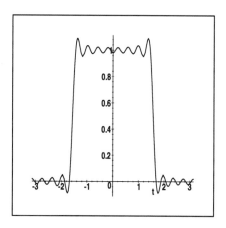

9.5   Laplace Transforms

Just as the Laplace transform

$$\mathcal{L}(f) = F(s) = \int_0^\infty f(t)e^{-st}\,dt$$

can facilitate the solution of ODEs, it is also useful with PDEs. Suppose u is now a function of two variables, and consider

$$\mathcal{L}(u(x,t)) = U(x,s) = \int_0^\infty u(x,t)e^{-st}\,dt.$$

If u is smooth enough, then

$$\mathcal{L}\left(\frac{\partial u}{\partial x}\right) = \int_0^\infty \frac{\partial u}{\partial x}(x,t)e^{-st}\,dt$$
$$= \frac{\partial}{\partial x}U(x,s) = \frac{dU}{dx}.$$

By the last symbolism, we simply announce our intention to keep the s variable in the background temporarily. As before,

$$\mathcal{L}\left(\frac{\partial u}{\partial t}\right) = sU(x,s) - u(x,0),$$

as one sees by integrating by parts.

Example

Consider the initial-boundary value problem

$$\frac{\partial^2 u}{\partial x^2} = \frac{\partial u}{\partial t}, \quad 0 < x < 1, \quad t > 0,$$
$$u(x,0) = 1 + \sin \pi x, \quad 0 < x < 1,$$
$$u(0,t) = 0, \quad u(1,t) = 0, \quad t > 0.$$

```
>  restart:with(inttrans):
>  alias(U(x,s)=laplace(u(x,t),t,s)):
>  diffeq:=diff(u(x,t),x,x)=diff(u(x,t),t);
```

$$diffeq := \frac{\partial^2}{\partial x^2}\, u(x,\,t) = \frac{\partial}{\partial t}\, u(x,\,t)$$

```
>  tfm:=laplace(diffeq,t,s);
```

$$tfm := \frac{\partial^2}{\partial x^2}\, U(x,\,s) = s\, U(x,\,s) - u(x,\,0)$$

```
>  subs(u(x,0)=1+sin(Pi*x),tfm);
```

$$\frac{\partial^2}{\partial x^2}\, U(x,\,s) = s\, U(x,\,s) - 1 - \sin(\pi\, x)$$

View s as fixed, so that U becomes a function of x only. We see this as an ordinary differential equation, which we ask Maple to solve.

```
>  ode:=diff(U(x),x,x)=s*U(x)-1-sin(Pi*x);
```

$$ode := \frac{\partial^2}{\partial x^2}\, U(x) = s\, U(x) - 1 - \sin(\pi\, x)$$

When $x = 0$ and when $x = 1$, we have $u(x, 0) = 1$, so the Laplace transform is $1/s$.

```
>  inits:=U(0)=1/s,U(1)=1/s;
```

$$inits := \mathrm{U}(0) = \frac{1}{s},\ \mathrm{U}(1) = \frac{1}{s}$$

```
>  psol:=rhs(dsolve({ode,inits},U(x)));
```

$$psol := \frac{s + \pi^2 + s\sin(\pi x)}{s\,(s + \pi^2)}$$

Note that $\sin \pi x$ is constant with respect to s, so we simply take the inverse Laplace transform.

```
>  invlaplace(psol,s,t);
```

$$1 + \sin(\pi x)\,e^{(-\pi^2 t)}$$

```
>  fsol:=(x,t)->1+sin(Pi*x)*exp(-Pi^2*t);
```

$$fsol := (x,\ t) \to 1 + \sin(\pi x)\,e^{(-\pi^2 t)}$$

```
>  plot3d(fsol(x,t),x=0..1,t=0..0.2);
```

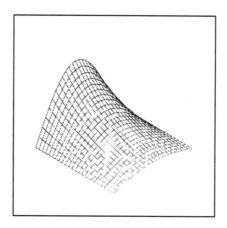

The reader can experiment with various view angles, axes, and styles.

Example

$$\frac{\partial^2 u}{\partial x^2} = \frac{\partial^2 u}{\partial t^2}, \quad 0 < x < 1, \quad t > 0,$$

$$u(x,0) = \sin \pi x, \quad \frac{\partial u}{\partial t}(x,0) = -\sin \pi x, \quad 0 < x < 1,$$

$$u(0,t) = 0, \quad u(\pi,t) = 0, \quad t > 0.$$

```
>  restart:with(inttrans):
```

```
>  alias(U(x,s)=laplace(u(x,t),t,s)):
```

```
>  diffeq:=diff(u(x,t),x,x)=diff(u(x,t),t,t);
```

$$diffeq := \frac{\partial^2}{\partial x^2} \, u(x, \, t) = \frac{\partial^2}{\partial t^2} \, u(x, \, t)$$

```
>  laplace(diffeq,t,s);
```

$$\frac{\partial^2}{\partial x^2} \, U(x, \, s) = s \, (s \, U(x, \, s) - u(x, \, 0)) - D_2(u)(x, \, 0)$$

Again, holding s constant, we get an ordinary differential equation. (Recall that the notation D_2 indicates the derivative with respect to the second variable, i.e., $\dfrac{\partial u}{\partial t}$.)

```
>  ode:=diff(U(x),x$2)=s*(s*U(x)-sin(Pi*x))+sin(Pi*x);
```

$$ode := \frac{\partial^2}{\partial x^2} \, U(x) = s \, (s \, U(x) - \sin(\pi \, x)) + \sin(\pi \, x)$$

Note that $u(0, t) = u(1, t) = 0$, so that the Laplace transforms are also both 0.

```
>  init:=U(0)=0,U(1)=0;
```

$$init := U(0) = 0, \, U(1) = 0$$

```
>  dsolve({ode,init},U(x));
```

$$U(x) = \frac{\sin(\pi \, x) \, s - \sin(\pi \, x)}{s^2 + \pi^2}$$

```
>  invlaplace(rhs("),s,t);
```

$$\sin(\pi \, x) \cos(\pi \, t) - \frac{\sin(\pi \, x) \sin(\pi \, t)}{\pi}$$

```
>  tt:=collect(",sin(Pi*x));
```

$$tt := \sin(\pi \, x) \left(\cos(\pi \, t) - \frac{\sin(\pi \, t)}{\pi} \right)$$

```
>  wave:=(x,t)->tt;
```

$$wave := (x, \, t) \rightarrow tt$$

```
>  plots[animate](wave(x,t),x=0..1,t=0..2,frames=50);
```

This seems to be relatively painless, but not really new, since the problem is easily solved by separation of variables. We now look at a problem that can't be directly solved by separation of variables.

Example

A string is supported from below and lies motionless on the positive x-axis. At time $t = 0$ the support is removed. The end at $x = 0$ is fixed. Find the displacement of the string.
The initial-boundary value problem is

$$\frac{\partial^2 u}{\partial t^2} = c^2 \frac{\partial^2 u}{\partial x^2} + g, \quad x > 0, \quad t > 0,$$

$$u(x,0) = 0, \quad \frac{\partial u}{\partial t}(x,0) = 0, \quad x > 0,$$

$$u(0,t) = 0, \quad t > 0,$$

where we take $g < 0$ to keep the "up" in the coordinate system being positive.

```
>   restart;with(inttrans):
```

```
>   alias(U(x,t)=laplace(u(x,t),t,s)):
```

```
>   pde:=diff(u(x,t),t,t)=c^2*diff(u(x,t),x,x)+g;
```

$$pde := \frac{\partial^2}{\partial t^2} \, \mathrm{u}(x,\, t) = c^2 \, (\frac{\partial^2}{\partial x^2} \, \mathrm{u}(x,\, t)) + g$$

```
>   laplace(pde,t,s);
```

$$s \, (s \, \mathrm{U}(x,\, t) - \mathrm{u}(x,\, 0)) - D_2(u)(x,\, 0) = c^2 \, (\frac{\partial^2}{\partial x^2} \, \mathrm{U}(x,\, t)) + \frac{g}{s}$$

Holding t constant, we get an ODE. We can then insert the conditions on the boundary and solve with dsolve.

```
>   ode:=(s*(s*U(x)))=c^2*diff(U(x),x,x)+g/s;
```

$$ode := s^2 \, \mathrm{U}(x) = c^2 \, (\frac{\partial^2}{\partial x^2} \, \mathrm{U}(x)) + \frac{g}{s}$$

```
>   dsolve(ode,U(x));
```

$$\mathrm{U}(x) = \frac{g}{s^3} + \_C1 \, e^{(\frac{s\,x}{c})} + \_C2 \, e^{(-\frac{s\,x}{c})}$$

Because the solution remains bounded as $x \to \infty$, the first constant must be zero. The boundary condition $u(0,t) = 0$, $t > 0$, requires that the second constant be $-g/s^3$.

```
>   U(x,s):=g/s^3*(1-exp(-s*x/c));
```

$$\mathrm{U}(x,\, s) := \frac{g \, (1 - e^{(-\frac{s\,x}{c})})}{s^3}$$

In order to plot solutions, assume $g = -10$ and plot as the x-coordinate the quantity x/c:

```
>   subs({c=1,g=-10},U(x,s));
```

$$-10 \, \frac{1 - e^{(-s\,x)}}{s^3}$$

Maple is unable to take the inverse Laplace transform until it knows the sign of x.

```
> assume(x,positive):invlaplace(",s,t);
```

$$-5\,t^2 + 5\,\text{Heaviside}(t - x\tilde{\ })\,t^2 - 10\,\text{Heaviside}(t - x\tilde{\ })\,t\,x\tilde{\ } + 5\,\text{Heaviside}(t - x\tilde{\ })\,x^{\tilde{\ }2}$$

```
> collect(",Heaviside);
```

$$(5\,t^2 - 10\,t\,x\tilde{\ } + 5\,x^{\tilde{\ }2})\,\text{Heaviside}(t - x\tilde{\ }) - 5\,t^2$$

To make the output more attractive, we substitute the $x\tilde{\ }$ by x with delayed evaluation single quotes in order to have Maple write the answer in terms of x. Then we remove the assumption on x. The unapply then converts the expression into a Maple function of two variables.

```
> subs(x='x',"):solf:=unapply(",x,t);x:='x':
```

$$solf := (x,\,t) \rightarrow (5\,t^2 - 10\,t\,x + 5\,x^2)\,\text{Heaviside}(t - x) - 5\,t^2$$

```
> plot([solf(x,p)$p=1..5],x=0..6);
```

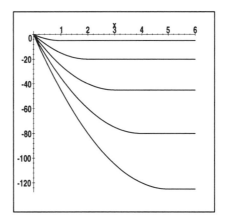

The string thus falls freely under gravity for large x/c until $t = x/c$. The shape of the string at time t_0 is parabolic until $x = ct_0$, and horizontal to the right of this point.

This problem cannot be solved using Fourier transform methods because the constant force of gravity is not integrable over $x > 0$.

9.6 Numerical Methods

Methods discussed so far work only in a tiny fraction of physically interesting boundary-initial value problems. Engineering problems are rarely set in rectangular regions with precisely known boundary and initial conditions, so numerical methods are needed. In this section we examine such a method.

We consider three examples: the heat equation, the wave equation, and Laplace's equation. In each case, we model derivatives by differences. Here are the replacements.

$$\frac{\partial u}{\partial x}(x,t) \rightarrow \frac{u(x+\Delta x, t) - u(x - \Delta x, t)}{2\Delta x}$$

$$\frac{\partial^2 u}{\partial x^2}(x,t) \rightarrow \frac{u(x+\Delta x, t) - 2u(x,t) + u(x - \Delta x, t)}{(\Delta x)^2}$$

$$\Delta u(x,y) = \frac{\partial^2 u}{\partial x^2}(x,y) + \frac{\partial^2 u}{\partial y^2}(x,y) \rightarrow$$

$$\frac{u(x+\Delta x, y) + u(x - \Delta x, y) + u(x, y + \Delta x) + u(x, y - \Delta x) - 4u(x,y)}{(\Delta x)^2}$$

At an end point of an interval $[a,b]$, say at $x = b$, we use

$$\frac{\partial u}{\partial x}(b,t) \rightarrow \frac{u(b,t) - u(b - \Delta x, t)}{\Delta x}.$$

This is sufficient to illustratrate the method in the heat equation.

A fundamental difference between spatial variables and time is simple: time "goes forward." For the heat equation,

$$\frac{\partial u}{\partial t}(x,t) \rightarrow \frac{u(x, t + \Delta t) - u(x,t)}{\Delta t}.$$

However, for the wave equation we will use

$$\frac{\partial^2 u}{\partial t^2}(x,t) \rightarrow \frac{u(x, t + \Delta t) - 2u(x,t) + u(x, t - \Delta t)}{(\Delta t)^2}.$$

In addition, we need to approximate the central difference,

$$\frac{\partial u}{\partial t}(x,t) \rightarrow \frac{u(x, t + \Delta t) - u(x, t - \Delta t)}{2\Delta t}$$

for the initial conditions.

We obtain two numerical approximations for each of the three equations. One will be like a boundary-initial value problem we have already solved; the other example cannot be solved using previous techniques.

9.6.1   Heat Equation

Example

Solve the initial-boundary value problem

$$\frac{\partial u}{\partial t}(x,t) = \frac{\partial^2 u}{\partial x^2}(x,t), \quad 0 < x < 1, \quad t > 0,$$
$$u(0,t) = u(1,t) = 0, \quad t \geq 0,$$
$$u(x,0) = \sin \pi x, \quad 0 \leq x \leq 1.$$

The exact solution is $u(x,t) = e^{-\pi^2 t} \sin \pi x$, as can be checked by direct substitution.

```
>  diffeq:=diff(u(x,t),x$2)-diff(u(x,t),t)=0;
```

$$diffeq := (\frac{\partial^2}{\partial x^2} \, u(x,t)) - (\frac{\partial}{\partial t} \, u(x,t)) = 0$$

```
> sol:=u(x,t)=exp(-Pi^2*t)*sin(Pi*x);
```

$$sol := \mathrm{u}(x,\, t) = e^{(-\pi^2 t)} \sin(\pi x)$$

```
> subs(sol,diffeq):simplify(");
```

$$0 = 0$$

For this illustration, first divide the interval $[0, 1]$ into n equal parts, making $\Delta x = \dfrac{1}{n}$. The time increment Δt will depend on Δx. There are $n + 1$ points along the x-axis, and many more t-points. Denote by $u_i(m)$ the temperature at location $x_i = i/n$ and at time $m\Delta t$. The PDE becomes

$$\frac{u_{i+1}(m) - 2u_i(m) + u_{i-1}(m)}{(\Delta x)^2} = \frac{u_i(m + 1) - u_i(m)}{\Delta t}.$$

Let $r = \Delta t/(\Delta x)^2$, and solve the difference equation for $u_i(m + 1)$:

$$u_i(m + 1) = ru_{i-1}(m) + (1 - 2r)u_i(m) + ru_{i+1}(m).$$

Here is the method: you know $u(x, 0) = \sin \pi x$. Evaluate the initial condition at the $n + 1$ points, giving $u_i(0)$. The formula gives $u_i(0 + 1)$, and so on. Boundary conditions in this problem set $u_0(m)$ and $u_n(m)$ to zero at each step.

There is a technical reason related to stability why the coefficients in the $u_i(m + 1)$ formula must not be negative. Therefore, the largest r can be is $1/2$. Using $1/2$ for r minimizes execution time, but a value of r nearer 0.3 produces a more accurate solution.

At this point, we digress briefly from the mathematics of partial differential equations to explain a little about the details of how these Maple routines were written. In particular, we have elected to use Maple's evalhf command, which enables the user to do double precision arithmetic (about 15 decimal digits of accuracy, normally) much faster than using Maple's software commands. In particular, the time to run all the commands in the section was cut in half by using evalhf. The reader who wishes to write his own routines is advised to study the Help pages, since if the command is improperly used, evalhf can even cause a 20% decrease in speed. The penalty arises from the fact that Maple stores all numbers internally in decimal form, whereas most computers store them as binary numbers. Converting back and forth from the decimal format to binary format is time consuming. As a consequence, the code needs to be structured so that most of the arithmetic is done in a single call to evalhf. Since normal passing of output from one subroutine to another forces Maple to convert forms, it is important to use the Maple command var, which passes arrays from subroutine to subroutine within a single call to evalhf, without any translation taking place. It may also be possible to save time by avoiding initialization of array entries to zero, since evalhf assumes that an undefined entry is zero. Consult the online help and *Maple V Programming Guide* by Monagan, Geddes, Labahn, and Verkoetter for further information.

```
> restart:
```

First we make an arrow defined function to calculate an individual entry in the array h.

```
> F:=(i,j,r,h)->evalf(r*(h[i-1,j-1]+h[i+1,j-1])+(1-2*r)*h[i,j-1]):
```

Then we write a procedure to give the entire array.

```
> nxt:=proc(nx,nt,dx,r,h)
  local i,j;
```

```
for i from 1 to nx-1 do
h[i,0]:=evalf(sin(Pi*i*dx)):
od;
for j from 1 to nt do
for i from 1 to nx-1 do
h[i,j]:=F(i,j,r,h);
od:
od:
end:
```

Finally, we write the procedure to approximate the solution. Here nx is the number of x points, $tmax$ is the largest value of t, r is the dt parameter, and h is the array of solution values.

```
>   nheat:=proc(nx,tmax,r)
    local i,j,nt;
    global dx,dt,h;
    dx:=evalhf(1/nx);
    dt:=evalhf(r*dx^2);
    nt:=ceil(tmax/dt);
    h:=array(0..nx,0..nt);
    print(nx,nt); # print num of x, t points
    evalhf(nxt(nx,nt,dx,r,var(h)));
    end:
```

```
>   nx:=16:tmax:=0.25:r:=0.3:
```

```
>   nheat(nx,tmax,r):
```

$$16, 214$$

The number 16, nx in the program, is the number of x-grid intervals; and 214, the global variable nt, is the number of time intervals. The parameter $tmax$ is the time limit.

```
>   with(plots):
```

```
>   phs:=(x,t)->h[round(x/dx),round(t/dt)]:
```

```
>   np:=numpoints=(nx+1)^2:
```

```
>   plot3d(phs,0..1,0..0.25,np);
```

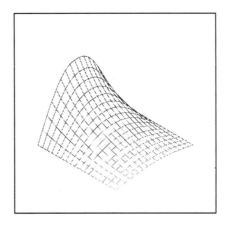

```
>   soln:=(x,t)->sin(Pi*x)*exp(-Pi^2*t):

>   np:=numpoints=(nx+1)^2:

>   plot3d(soln,0..1,0..0.25,np);
```

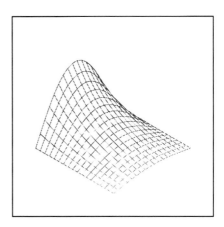

```
>   errr:=soln(x,t)-phs(x,t):

>   np:=numpoints=(nx+1)^2:

>   plot3d(errr,x=0..1,t=0..0.25,np);
```

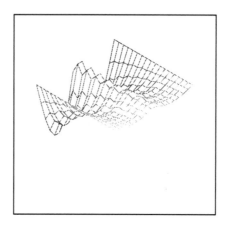

By using the coordinate frame buttons to rotate the graph and by inserting coordinate axes, you can see that the error is between −0.033 and 0. Although there is some error owing to making the problem discrete, the main source of error is in the Euler method approximation to the time derivative. Unfortunately, increasing the number of points by, say, doubling *nx* will increase the time taken by the algorithm by a factor of eight.

Example

Solve the initial-boundary value problem

$$\frac{\partial u}{\partial t}(x,t) = \frac{\partial^2 u}{\partial x^2}(x,t), \quad 0 < x < 1, \quad t > 0,$$
$$u(0,t) = 2t, \quad \frac{\partial u}{\partial x}(1,t) = 0, \quad t \geq 0,$$
$$u(x,0) = 2\sin \pi x, \quad 0 \leq x \leq 1/2,$$
$$u(x,0) = 1, \quad 1/2 < x \leq 1.$$

In this problem, the heat conducting rod is insulated at the $x = 1$ end, is held at temperature $2t$ at the $x = 0$ end, and is subjected to a slightly different initial condition. It is easy to modify the above code to handle this problem.

```
>   restart:

>   F:=(i,j,r,h)->evalf(r*(h[i-1,j-1]+h[i+1,j-1])+(1-2*r)*h[i,j-1]):

>   nxt1:=proc(nx,nt,dx,r,h)
    local i,j;
    for i from 1 to nx-1 do
    if i<nx/2 then
    h[i,0]:=evalf(2*sin(Pi*i*dx))
    else h[i,0]:=1.0 fi:
    od;
    for j from 1 to nt do
    h[0,j]:=2*j*dt;
    for i from 1 to nx-1 do
```

```
        h[i,j]:=F(i,j,r,h);
        od:
        od:
        end:
>   nheat:=proc(nx,tmax,r)
        local i,j,nt;
        global dx,dt,h;
        dx:=evalf(1/nx);
        dt:=evalf(r*dx^2);
        nt:=ceil(tmax/dt);
        h:=array(0..nx,0..nt);
        print(nx,nt); # print num of x, t points
        evalhf(nxt1(nx,nt,dx,r,var(h)));
        end:
>   nx:=16:tmax:=0.25:r:=0.3:
>   nheat(nx,tmax,r):
```

$$16, 214$$

```
>   phs:=(x,t)->h[round(x/dx),round(t/dt)]:
>   with(plots):
>   np:=numpoints=(nx+1)^2:
>   plot3d(phs,0..1,0..0.25,np);
```

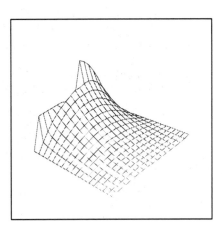

This problem could not have been solved directly by the techniques discussed to this point.

9.6.2 Wave Equation

The wave equation we consider now is

$$\frac{\partial^2 u}{\partial x^2} = \frac{\partial^2 u}{\partial t^2}.$$

With spatial step size Δx and a time step Δt, this translates into the discrete equation

$$\frac{u(x + \Delta x, t) - 2u(x,t) + u(x - \Delta x, t)}{(\Delta x)^2} = \frac{u(x, t + \Delta t) - 2u(x,t) + u(x, t - \Delta t)}{(\Delta t)^2}.$$

Using the same convention as before, at the ith point along the string and the mth time point, the equation is

$$\frac{u_{i+1}(m) - 2u_i(m) + u_{i-1}(m)}{(\Delta x)^2} = \frac{u_i(m + 1) - 2u_i(m) + u_i(m - 1)}{(\Delta t)^2}.$$

The parameter r (for the heat equation) is now $\rho = (\Delta t/\Delta x)^2$. Solve for $u_i(m + 1)$:

$$u_i(m + 1) = \rho u_{i-1}(m) + 2(1 - \rho)u_i(m) + \rho u_{i+1}(m) - u_i(m - 1).$$

The numerical solution of the wave equation must use the end-point conditions, as well as the initial position and velocity of the string. For example:

$$u(0, t) = u(1, t) = 0, \quad t \geq 0,$$
$$u(x, 0) = f(x), \quad 0 \leq x \leq 1,$$
$$\frac{\partial u}{\partial t}(x, 0) = g(x), \quad 0 < x < 1.$$

For the wave equation, we need the central difference approximation to the first derivative:

$$\frac{\partial u}{\partial t}(x, 0) \rightarrow \frac{u(x, \Delta t) - u(x, -\Delta t)}{2\Delta t}.$$

However, this approach would require us to compute $u_i(-1)$, so we combine the first step equation and the boundary condition to eliminate $u_i(-1)$:

$$u_i(1) + u_i(-1) = \rho f(x_{i-1}) + 2(1 - \rho)f(x_i) + \rho f(x_{i+1}),$$
$$u_i(1) - u_i(-1) = 2\Delta t g(x_i).$$

Adding these equations yields:

$$u_i(1) = \frac{1}{2}\rho \left(f(x_{i-1}) + f(x_{i+1}) \right) + (1 - \rho)f(x_{i+1}) + \Delta t\, g(x_i).$$

(We could have used the same strategy in the heat equation with an insulated end, but the numerical result would not have been significantly different. For the wave equation, the initial conditions are much more critical.)

Example

Return to the example of the string plucked in the center by displacing it by 0.1 and releasing. Note that g, the initial velocity in the original problem, is zero, a fact which would allow us to simplify the Maple code; but we will keep the structure of the initialization for the next example.

```
>  restart:

>  F:=(i,j,r,w)->evalf(r*(w[i-1,j-1]+w[i+1,j-1])+2*(1-r)*w[i,j-1]-w[i,j-2]):
```

```
>   nxt2:=proc(nx,nt,dx,r,w)
    local i,j,f;
    f:=id->if 2*id<nx then 0.2*id*dx
    else 0.2*(nx-id)*dx fi;
    for i from 1 to nx-1 do
    w[i,0]:=evalhf(f(i));
    od;
    for i from 1 to nx-1 do
    w[i,1]:=evalf(r/2*(f(i-1)+f(i+1))
    +(1-r)*f(i));
    od;
    for j from 2 to nt do
    for i from 1 to nx-1 do
    w[i,j]:=F(i,j,r,w);
    od;
    od:
    end:


>   waveeqn:=proc(nx,tmax,r)
    local i,j,f,id;
    global dx,dt,w,nt;
    dx:=evalf(1/nx);
    dt:=evalf(sqrt(r)*dx);
    nt:=ceil(tmax/dt);
    print(nx,nt);
    w:=array(0..nx,0..nt);
    evalhf(nxt2(nx,nt,dx,r,var(w)));
    end:


>   nx:=32:tmax:=2:rho:=1:


>   waveeqn(nx,tmax,rho);
```

$$32, 64$$

$$.006249999999999990$$

```
>   with(plots):


>   ps:=(x,t)->w[round(x/dx),round(t/dt)]:


>   np:=numpoints=(nt+1)^2:


>   plot3d(ps,0..1,0..tmax,np);
```

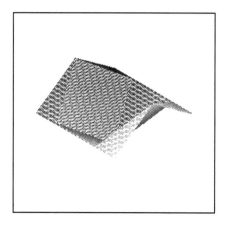

```
>  fr:=frames=tmax*16:

>  animate(ps(x,t),x=0..1,t=0..tmax,fr);
```

Two features are worth noting: first, the period of one complete cycle is 2, just as in the case of the analytical solution. Second, there is no loss, in the sense that the numerical solution returned to the same position as it started in one cycle. (This would not be as precisely true if the initial velocity were not zero.)

Example

This example models a vibrating string with a mechanical resistance term proportional to the velocity. There are advanced methods for solving such equations, but they are far beyond the scope of this manual. Recall that Maple *can* solve the wave equation. Before we continue, we check whether Maple can handle this problem, which is "close" to the wave equation.

```
>  restart:

>  infolevel[pdesolve]:=9:

>  pde:=diff(u(x,t),x,x)=diff(u(x,t),t,t)+diff(u(x,t),t);
```

$$pde := \frac{\partial^2}{\partial x^2}\, \mathrm{u}(x,\,t) = (\frac{\partial^2}{\partial t^2}\, \mathrm{u}(x,\,t)) + (\frac{\partial}{\partial t}\, \mathrm{u}(x,\,t))$$

```
>  pdesolve(pde,u(x,t));
```

```
pdesolve/analyze:    equation order    2
pdesolve/analyze:    derivatives in equation:    \{t, x\}
pdesolve/analyze:    not an algebraic equation
pdesolve/analyze:    not an ode in obvious disguise
pdesolve/analyze:    not homogeneous monomial equation
pdesolve/analyze:    linear equation
pdesolve/analyze:    not reducible lin. const. coeff.
pdesolve/analyze:    not lin. const. coeff. 1st order
```

```
pdesolve/analyze:   trying algorithms for non-constant coeffs
pdesolve/analyze:   not reducible lin. var. coeff, 1st order
pdesolve/analyze:   not in a 'linear' subset of Lagrange
pdesolve/analyze:   trying to factor (non-commuting) pde
pdesolve/analyze:   cannot factor linear pde in the plane
pdesolve/analyze:   trying algorithms for non-linear
pdesolve/analyze:   not first order equation
pdesolve/analyze:   do not know how to handle equation
pdesolve/exact/2:   Do not know how to handle equation
```

$$\text{pdesolve}(\frac{\partial^2}{\partial x^2}\,\text{u}(x,\,t) = (\frac{\partial^2}{\partial t^2}\,\text{u}(x,\,t)) + (\frac{\partial}{\partial t}\,\text{u}(x,\,t)),\,\text{u}(x,\,t))$$

Since Maple cannot solve it directly, numerical methods are appropriate.

The initial-boundary value problem that we now consider has the same initial-boundary values as the previous example, but the PDE is modified by adding a damping term:

$$\frac{\partial^2 u}{\partial x^2} = \frac{\partial^2 u}{\partial t^2} + b\frac{\partial u}{\partial t}.$$

When we construct the approximating differences, the starting equations remain the same as before, since the damping term is proportional to velocity, which is initially zero. In contrast, the running equation, the equation at time step $m > 1$, is changed. Recall that the time first derivative is modeled by the central first difference:

$$\frac{\partial u}{\partial t}(x,t) \rightarrow \frac{u(x,t+\Delta t) - u(x,t-\Delta t)}{2\Delta t}$$

In our discrete notation, this is $(u_i(m+1) - u_i(m-1))/(2\,dt)$. This modification then contributes another term to the expression for $u_i(m+1)$:

$$u_i(m+1) = \rho u_{i-1}(m) + 2(1-\rho)u_i(m) + \rho u_{i+1}(m) - u_i(m-1) - b\Delta t(u_i(m+1) - u_i(m-1))/2.$$

Solve this for $u_i(m+1)$:

$$(1 + \frac{b\Delta t}{2})u_i(m+1) = \rho u_{i-1}(m) + 2(1-\rho)u_i(m) + \rho u_{i+1}(m) - (1 - \frac{b\Delta t}{2})u_i(m-1).$$

```
>   restart:

>   F:=(i,j,r,w,f1,f2,f3)->evalf(f1*(w[i-1,j-1]+w[i+1,j-1])+f2*w[i,j-1]-f3*w[i,j-2])

>   nxt3:=proc(nx,nt,dx,r,w,f1,f2,f3)
    local i,j,f;
    f:=id->if 2*id<nx then
    evalf(0.2*id*dx)
    else evalf(0.2*(nx-id)*dx) fi;
    for i from 1 to nx-1 do
    w[i,0]:=evalhf(f(i));
    od;
    for i from 1 to nx-1 do
    w[i,1]:=evalhf(r/2*(f(i-1)+f(i+1))
    +(1-r)*f(i));
    od;
```

```
        for j from 2 to nt do
        for i from 1 to nx-1 do
        w[i,j]:=F(i,j,r,w,f1,f2,f3);
        od;
        od:
        end:
```

```
>    waveeqn:=proc(nx,tmax,rho,b)
     local i,j,f,id,f1,f2,f3;
     global dx,dt,w,nt;
     dx:=evalf(1/nx);
     dt:=evalf(sqrt(rho)*dx);
     nt:=ceil(tmax/dt);
     f1:=evalf(rho/(1+b*dt/2));
     f2:=evalf(2*(1-rho)/(1+b*dt/2));
     f3:=evalf((1-b*dt/2)/(1+b*dt/2));
     print(nx,nt);
     w:=array(0..nx,0..nt);
     evalhf(nxt3(nx,nt,dx,rho,var(w),f1,f2,f3));
     end:
```

```
>    nx:=32:tmax:=8:rho:=1:b:=0.25:
```

```
>    waveeqn(nx,tmax,rho,b);
```

$$32, 256$$

$$.002296447008680340$$

```
>    ps:=(x,t)->w[round(x/dx),round(t/dt)]:
```

```
>    np:=numpoints=(nt+1)*(nx+1):
```

```
>    plot3d(ps,0..1,0..tmax,np);
```

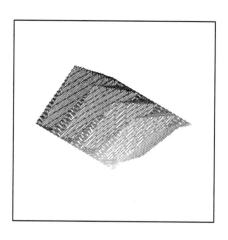

The three dimensional plot of the solution is not as easy to understand as an animation:

```
>  with(plots):
>  fr:=frames=16*tmax:
>  animate(ps(x,t),x=0..1,t=0..tmax);
```

The center portion of the string is no longer flat, but "sags" toward the x-axis. This is because the velocity is higher there, and so the resistance term is relatively stronger. The period of the vibration does not change as more cycles are observed. The period remains $t = 2$, independent of the damping factor b. Contrast this to the behavior of the damped mass-spring problem, in which the period depends on the damping factor.

9.6.3   Laplace's Equation

Example

In this example, the region R is the unit square with u specified on the top and bottom, and the left and right boundaries insulated.

$$\frac{\partial^2 u}{\partial x^2} + \frac{\partial^2 u}{\partial y^2} = 0, \quad 0 < x < 1, \quad 0 < y < 1,$$

$$\frac{\partial u}{\partial x}(0, y) = \frac{\partial u}{\partial x}(1, y) = 0, \quad 0 \le y \le 1,$$

$$u(x, 1) = 0, \quad u(x, 0) = 1 - \cos(2\pi x), \quad 0 \le x \le 1.$$

This example was solved earlier in the rectangle $[0, \pi] \times [0, 1]$ using separation of variables. Recall that the solution was given as a series. To obtain the numerical value of the solution, one had to calculate the sum of the series, usually numerically. So why not solve the equation numerically from the beginning?

To this end, divide the x- and y-axes into n equal subdivisions of length $\Delta x = 1/n$. Denote by $u_{i,j}$ the solution at $x_i = i/n$ and $y_j = j/n$. Inside the square, the discrete Laplacian is

$$\Delta u_{i,j} = \frac{u_{i-1,j} - 2u_{i,j} + u_{i+1,j}}{(\Delta x)^2} + \frac{u_{i,j-1} - 2u_{i,j} + u_{i,j+1}}{(\Delta x)^2} = 0, \quad 0 < i < n, \quad 0 < j < n.$$

That is,

$$u_{i,j} = \frac{1}{4}(u_{i-1,j} + u_{i+1,j} + u_{i,j-1} + u_{i,j+1}).$$

For the boundary conditions, set the value to the prescribed value for Dirichlet boundary conditions, and to the value of the nearest neighbor inside the square for Neumann boundary conditions.

The resulting problem is a special kind of linear algebra problem. Except for the boundary condition equations, the non-zero matrix entries are clustered along the diagonal. One numerical method for obtaining a solution, known as the *Gauss-Seidel* method, takes an initial distribution of u-values (perhaps all zero) and uses the equations and boundary equations to replace the old values with new ones. In the process, the most recently calculated values are used. (This turns out to be more accurate and about twice as fast as using all of the old values at each step.) Of course, this seems to go on forever, so two methods are commonly used to stop the iteration. One is to stop when the maximum change from one step to the next is less than a prescribed tolerance (*tol* in the procedure which follows); another is to limit the maximum number of iterations (1000 here).

This is easily programmed in Maple:

```
> restart:

> F:=(i,j,l)->evalf((l[i-1,j]+l[i+1,j]+l[i,j-1]+l[i,j+1])/4):

> nxt4:=proc(n,l,tol) local i,j,it,test,new,ch;
  ch:=1;
  for it from 1 to 1000 while ch>tol do
  ch:=0;
  for i from 0 to n do
  l[i,0]:=evalf(1-cos(2*Pi*i/n)):
  od;
  for i from 1 to n-1 do
  for j from 1 to n-1 do
  new:=F(i,j,l);
  test:=evalf(abs(new-l[i,j]));
  if ch<test then ch:=test fi;
  l[i,j]:=new;
  od:
  od;
  for j from 1 to n do
  l[0,j]:=l[1,j];
  l[n,j]:=l[n-1,j];
  od;
  od;
  print(ch,it);
  end:

> laplace:=proc(n,tol)
  local i,j,ch;
  global l;
  l:=array(0..n,0..n);
  evalhf(nxt4(n,var(l),tol));
  end:

> n:=16:tol:=0.001:

> laplace(n,tol):
```

$$.0009830365084133597, 139$$

The two numbers printed indicate that the maximum change in the solution value on this iteration was about the specified tolerance 0.001 after 139 iterations.

```
> with(plots):

> pl:=(x,y)->l[round(x*n),round(y*n)]:

> np:=numpoints=(n+1)^2:

> plot3d(pl,0..1,0..1,np);
```

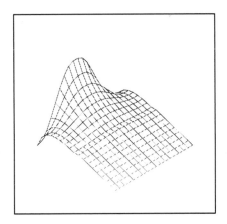

```
> fr:=frames=50:
> animate(pl(x,y),x=0..1,y=0..1,fr);
```

The solution could have been obtained faster in this example by solving the linear algebra problem using the tools in the linalg package; however, considerably more programming (and hence human) time would be needed to set up the problem.

The next example illustrates a natural problem that could not be solved using separation of variables.

Example

A circular pipe with diameter $1/2$ contains fluid at constant temperature 0. The pipe is centered in a 1×1 square containing insulation with constant heat conductivity. The temperature of the outside of the square is a constant 1. What is the steady state temperature distribution in the insulation?

A cross section of the pipe-square is a two dimensional Dirichlet problem.

$$\frac{\partial^2 u}{\partial x^2} + \frac{\partial^2 u}{\partial y^2} = 0, \quad 0 < x < 1, \quad 0 < y < 1, \quad \left(x - \frac{1}{2}\right)^2 + \left(y - \frac{1}{2}\right)^2 > \frac{1}{4},$$
$$u(x,0) = u(x,1) = 1, \quad 0 \le x \le 1,$$
$$u(0,y) = u(1,y) = 1, \quad 0 \le y \le 1,$$
$$u(x,y) = 0, \quad \left(x - \frac{1}{2}\right)^2 + \left(y - \frac{1}{2}\right)^2 = \tfrac{1}{4}.$$

The code which solves the previous example can be adapted to solve this problem; it is also clear how the code can be changed to account for more exotic boundary conditions.

```
> restart:
> F:=(i,j,l)->evalf((l[i-1,j]+l[i+1,j]+l[i,j-1]+l[i,j+1])/4):
```

```
>   nxt5:=proc(n,l,tol)
    local i,j,it,test,new,ch;
    ch:=1.0;
    for it from 1 to 1000 while ch>tol do
    ch:=0.0;
    for i from 1 to n-1 do
    for j from 1 to n-1 do
    if 4*((2*i-n)^2+(2*j-n)^2)>n^2 then
    new:=F(i,j,l);
    test:=evalf(abs(new-l[i,j]));
    if ch<test then ch:=test fi;
    l[i,j]:=new;
    fi;
    od:
    od;
    od;
    print(ch,it);
    end:

>   laplace:=proc(n,tol)
    local i,j,ch;
    global l;
    l:=array(0..n,0..n);
    for i from 0 to n do
    l[i,0]:=1.0:
    l[i,n]:=1.0;
    l[0,i]:=1.0;
    l[n,i]:=1.0.
    od;
    evalhf(nxt5(n,var(l),tol));
    end:

>   n:=32:tol:=.005:

>   laplace(n,tol):
```

$$.004772105363922530, 52$$

```
>   pl:=(x,y)->l[round(x*n),round(y*n)]:

>   np:=numpoints=(n+1)^2:

>   st:=style=patchcontour:

>   ct:=contours=10:

>   ort:=orientation=[45,-110]:

>   plot3d(pl,0..1,0..1,np,st,ct,ort);
```

Other views can be had by using the plot option buttons.
Here is a contour map of the temperature distribution.

```
>  with(plots):

>  con:=contours=[0.0,0.2,0.4,0.6,0.8,1.0]:

>  np:=numpoints=(n+1)^2:

>  contourplot(pl,0..1,0..1,con,np);
```

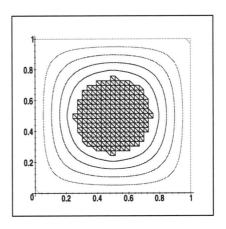

Finally, successive slices through the plane can be displayed as an animation.

```
>  np:=numpoints=(n+1):

>  fr:=frames=70:

>  animate(pl(x,y),x=0..1,y=0..1,np,fr);
```

9.7   Exercises

In each exercise you will use a combination of "hand" analysis and Maple to produce solutions. Often, the main contribution from Maple is graphical.

1. Find the Fourier series of the function given by

$$f(x) = \begin{cases} 0 & \text{if } -\pi < x \le 0, \\ x & \text{if } 0 < x \le \pi, \end{cases} \qquad f(x + 2\pi) = f(x), \quad \text{all } x.$$

   Sketch partial sums corresponding to several values of N.

2. Verify that the function

$$u(x, t) = \frac{1}{\sqrt{4\pi t}} \exp\left(-\frac{x^2}{4t}\right)$$

   is a solution to the heat equation

$$\frac{\partial^2 u}{\partial x^2} = \frac{\partial u}{\partial t}, \quad t > 0, \quad -\infty < x < \infty.$$

   Sketch $u(x, t)$ for various ranges of x and t.

3. Solve

$$\frac{\partial^2 u}{\partial x^2} = \frac{1}{\beta}\frac{\partial u}{\partial t}, \quad t > 0, \quad 0 < x < \pi,$$
$$u(0, t) = 0, \quad u(\pi, t) = T, \quad t > 0,$$
$$u(x, 0) = 0, \quad 0 < x < \pi.$$

4. One important problem in mathematical physics is the Dirichlet problem in a rectangle. Solve the equation in this example.

$$\frac{\partial^2 u}{\partial x^2} + \frac{\partial^2 u}{\partial y^2} = 0, \quad 0 < x < \pi, \quad 0 < y < 4,$$
$$u(x, 0) = u(x, 4) = \sin x, \quad 0 < x < \pi,$$
$$u(0, y) = u(\pi, y) = 0, \quad 0 < y < 4.$$

5. Solve the next three problems and compare the solutions. You might want to try other functions in addition to $1 - \cos 2x$. Pay particular attention to elements of the methods that are common to all three equations.

   (a) Heat equation:

$$\frac{\partial^2 u}{\partial x^2} = \frac{\partial u}{\partial y}, \quad 0 < x < \pi, \quad y > 0,$$
$$u(x, 0) = 1 - \cos 2x, \quad 0 < x < \pi,$$
$$u(0, y) = 0, \quad u(\pi, y) = 0, \quad y > 0.$$

(b) Wave equation:

$$\frac{\partial^2 u}{\partial x^2} = \frac{\partial^2 u}{\partial y^2}, \quad 0 < x < \pi, \ y > 0,$$

$$u(x,0) = 1 - \cos 2x, \quad \frac{\partial u}{\partial y}(x,0) = 0, \quad 0 < x < \pi,$$

$$u(0,y) = 0, \quad u(\pi,y) = 0, \quad y > 0.$$

(c) Laplace's equation:

$$\frac{\partial^2 u}{\partial x^2} + \frac{\partial^2 u}{\partial y^2} = 0, \quad 0 < x < \pi, \quad y > 0,$$

$$u(x,0) = 1 - \cos 2x, \quad 0 < x < \pi,$$

$$u(0,y) = 0, \quad u(\pi,y) = 0, \quad y > 0.$$

Chapter 10

Projects

This chapter contains a number of projects dealing with ordinary differential equations. These assignments are more involved than the exercises in the preceding chapters.

10.1   Heat-Seeking Particles

Objectives

The goal of this project is to determine the path of a heat-seeking particle moving in the plane or in space.

Background

An ideal heat-seeking particle is a particle that has the property that it always moves in the direction of maximal temperature increase. We are not concerned with the mechanics of such a particle other than its heat-seeking property: if it is located at point P, then it moves in the direction in which temperature is increasing most rapidly; i.e., it moves in the direction of the gradient of the temperature function evaluated at P.

The path of a heat-seeking particle in the plane

Suppose the temperature at the point (x, y) is given by the temperature function $T(x, y)$. We desire to determine the path of a heat-seeking particle that is initially located at the point (x_0, y_0). We know from multivariable calculus that at (x, y) the direction of maximal temperature increase is given by the gradient vector

$$\vec{\nabla} T(x, y) = \frac{\partial T}{\partial x}(x, y)\ \vec{\imath} + \frac{\partial T}{\partial x}(x, y)\ \vec{\jmath}.$$

Since our particle is heat-seeking, at the point (x, y) it moves in the direction of $\vec{\nabla} T(x, y)$. This means that $\vec{\nabla} T(x, y)$ is tangent to the path of our particle, so at every point on the particle's path we have a tangent vector to the path. (This is analogous to having, at each point on the curve, a tangent line to a curve.) Since we are only interested in the path of the particle we may assume that $\vec{\nabla} T(x, y)$ is the particle's velocity vector at (x, y). If $(x(t), y(t))$ gives the location at time t of the particle, then

$$\frac{dx}{dt}\ \vec{\imath} + \frac{dy}{dt}\ \vec{\jmath} = \vec{\nabla} T(x, y)$$

215

or

$$\begin{cases} \dfrac{dx}{dt} = \dfrac{\partial T}{\partial x}, \\[2mm] \dfrac{dy}{dt} = \dfrac{\partial T}{\partial y}, \end{cases} \tag{10.1}$$

with the initial conditions $x(0) = x_0$, $y(0) = y_0$.

Instructions

1. Show that if (10.1) turns out to be a pair of linear differential equations with constant coefficients

$$\begin{cases} \dfrac{dx}{dt} = \alpha x + \beta y, \\[2mm] \dfrac{dy}{dt} = \gamma x + \delta y, \end{cases}$$

then $\gamma = \beta$; i.e., the matrix of coefficients is symmetric. (This result is reasonable since our system is giving us the path of a heat-seeking particle. Our particle's path cannot oscillate due to its heat-seeking nature. Recall that symmetric matrices have real eigenvalues and so the solutions to the corresponding system will not oscillate.)

2. (a) Find the path of a heat-seeking particle that is initially at the point $(1, 2)$ if the temperature in the xy-plane is given by the temperature function $T(x, y) = xy$.

   (b) Notice that at $(0, 0)$ we have $\vec{\nabla} T(x, y) = \vec{0}$ so at $(0, 0)$ the rate of change of temperature is 0 in every direction. Will the particle that is initially at $(1, 2)$ stop at $(0, 0)$? Does this agree with your intuition?

   (c) Where would a heat-seeking particle have to be initially in order for it to go to $(0, 0)$?

3. Suppose the temperature at (x, y) is given by $T(x, y) = 4xy - x^4 - y^4 - 1$.

   (a) Use Maple to draw a graph of a heat-seeking particle that is initially at $(2, 5)$. Where does this particle end up?

   (b) Same as (3a), but the particle starts at $(-2, -5)$.

The path of a heat-seeking particle in space

This is a straightforward generalization of our analysis of motion in the plane. In space we will have a temperature function $T(x, y, z)$ and the system of differential equations that describes the path $(x(t), y(t), z(t))$ of our heat-seeking particle initially at the point (x_0, y_0, z_0) is

$$\begin{cases} \dfrac{dx}{dt} = \dfrac{\partial T}{\partial x}, \\[2mm] \dfrac{dy}{dt} = \dfrac{\partial T}{\partial y}, \\[2mm] \dfrac{dz}{dt} = \dfrac{\partial T}{\partial z}, \end{cases}$$

with initial conditions $x(0) = x_0$, $y(0) = y_0$, $z(0) = z_0$.

Instructions

Find the path of a heat-seeking particle that is initially at $(1, 2, 5)$ if the temperature in space is given by $T(x, y, z) = xy + xz + yz$.

10.2 Autonomous Differential Equations

Objective

This project investigates the behavior of solutions to an autonomous differential equation without actually solving the equation.

Background: first order autonomous differential equations

A first order autonomous differential equation is a differential equation of the form $x' = f(x)$, where the independent variable t does not explicitly appear in the equation. (We are using $x(t)$ as our unknown function of t instead of $y(t)$. The reason for this should be clear shortly.) Examples of first order autonomous differential equations are $x' = x^2 + x + 1$, $x' = \cos x$, $x' = e^x + 10$, etc. The differential equation $x' = x^2 + 2t$ is not autonomous since t appears explicitly in the equation.

Note that every first order autonomous differential equation is a separable differential equation and in theory could be solved by the technique of separation of variables.

Motion of a particle along the *x*-axis

We may regard the function $x(t)$ as giving the location at time t of a particle moving along the x-axis. With this in mind $x'(t)$ is the velocity of the particle at time t. (The particle is moving to the right on the x-axis if $x'(t)$ is positive and to the left if $x'(t)$ is negative.) The differential equation $x' = f(x)$ relates the position of the particle to its velocity.

As an example consider the equation $x'(t) = (x(t) - 1)(x(t) - 2)$, or more briefly $x' = (x - 1)(x - 2)$. We imagine a particle moving along the x-axis, its position at time t is $x(t)$ and its velocity at time t is $x'(t) = (x(t) - 1)(x(t) - 2)$. We want to know how the particle moves if its position at time 0 is given by $x(0) = x_0$.

Using Maple to solve $x' = (x - 1)(x - 2)$ gives us the general solution.

```
>  diffeq:=diff(x(t),t)=(x(t)-1)*(x(t)-2);
```

$$diffeq := \frac{\partial}{\partial t} x(t) = (x(t) - 1)(x(t) - 2)$$

```
>  sol:=dsolve(diffeq,x(t));
```

$$sol := \ln(x(t) - 1) - \ln(x(t) - 2) + t = \_C1$$

This does not seem to be very helpful in picturing how the particle moves along the x-axis especially since $x(t)$ is described implicitly in terms of t. So let's try assigning our particle an initial position of $x(0) = 1/2$ and have Maple solve the corresponding initial value problem to find the path of our particle:

$$x' = (x - 1)(x - 2), \quad x(0) = \frac{1}{2}.$$

Maple gives:

```
>   inits:=x(0)=1/2;
```

$$inits := x(0) = \frac{1}{2}$$

```
>   dsolve({diffeq,inits},x(t)):sol:=simplify(",exp);
```

$$sol := x(t) = \frac{-2 + 3\,e^t}{3\,e^t - 1} \tag{10.2}$$

This is better. At least the position of the particle at time t has been found explicitly. Still something is lacking since this solution does not give one a clear feeling about the path of the particle.

There is a simpler way to discover the path of our moving particle. Continuing with the above example, we see that $x' = 0$ precisely when $x = 1$ and $x = 2$. At these two locations our particle will have velocity 0. Moveover, $x' > 0$ when $x < 1$ and when $x > 2$. This says that at these locations the particle will be moving to the right. Finally, $x' < 0$ when $1 < x < 2$, and then the particle is moving to the left. The above information may be readily obtained by having Maple plot the function $f(x) = (x - 1)(x - 2)$ on an x, x'-coordinate system.

```
>   plot((x-1)*(x-2),x=-1..4,y=-1..3,labels=['x','x'']);
```

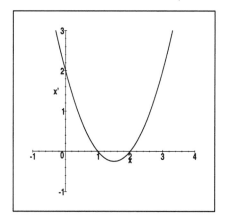

The graph is 0 at $x = 1$ and $x = 2$, positive on $(-\infty, 1)$ and on $(2, \infty)$ and negative on $(1, 2)$.

Now select a point x_0 on the x-axis between $x = 1$ and $x = 2$. If our particle is at x_0 its velocity is negative and so it will move to the left toward the point $x = 1$. When it reaches $x = 1$ it will have zero velocity and stop. (In fact, the particle will not reach the point $x = 1$ because this point is a limiting point, $\lim_{t\to\infty} x(t) = 1$. But since we are interested only in the path of the particle, this technicality need not concern us.) A particle initially located at x_0 where $x_0 < 1$ (as with the case $x_0 = 1/2$ above) will have positive velocity and move to $x = 1$ and stop. A particle located at x_0, where $x_0 > 2$, will move away from the point $x = 2$ indefinitely. This can be summarized by putting arrows on the x-axis indicating the direction of motion to obtain a "flow diagram" as pictured.

So the path of the particle initially at $x_0 = 1/2$ given by (10.2) is especially simple. It starts at $x_0 = 1/2$, moves to $x = 1$, and stops—it can go no further and cannot return.

The points $x = 1$ and $x = 2$ where the particle has zero velocity are called *fixed points*. Note that our example has two kinds of fixed points: *attractors* ($x = 1$) and *repellers* ($x = 2$).

Instructions

1. Have Maple solve the initial value problem $x' = (x-1)(x-2)$, $x(0) = 0$. Compute $\lim_{t\to\infty} x(t)$. Does this answer agree with the above analysis?

2. For each of the autonomous equations $x' = f(x)$ given below, do the following:

   (i) Find all fixed points.

   (ii) Classify each fixed point as an attractor or a repeller.

   (iii) Draw a flow diagram similar to that above.

   (a) $x' = 1 - x^2$

   (b) $x' = 1 + x + x^2$

   (c) $x' = \sin(x)$

   (d) $x' = 1 - x^{14}$

   (e) $x' = e^x - \cos(x)$

3. Explain intuitively why a solution to $x' = f(x)$ cannot "oscillate."

Systems of first order autonomous differential equations

Suppose that we have a particle moving in the xy-plane. Let the location of the particle at time t be the point $(x(t), y(t))$. Then the velocity of the particle at time t is $(x'(t), y'(t))$. If the velocity of the particle is a function of position only then we have a system of first order autonomous differential equations

$$x' = f(x, y),$$
$$y' = g(x, y). \tag{10.3}$$

If the particle is located at the point (x_0, y_0) at time $t = 0$, then we have the initial value problem

$$x' = f(x, y), \quad y' = g(x, y),$$
$$x(0) = x_0, \quad y(0) = y_0.$$

The analysis of solutions to (10.3) is more complicated than that for a single equation because of the freedom of movement in the xy-plane as opposed to movement along the x-axis. The analysis of solutions to linear autonomous systems

$$x' = \alpha x + \beta y,$$
$$y' = \gamma x + \delta y,$$

is completely discussed in your text.

Instructions

Assume the location of a moving particle in the xy-plane is given by the linear system (6.1). Let $A = \begin{bmatrix} \alpha & \beta \\ \gamma & \delta \end{bmatrix}$.

1. Show that if A is invertible, then the only point in the xy-plane at which a particle has zero velocity is the origin $(0, 0)$.

2. If A is not invertible and not the zero matrix, then there are infinitely many points in the xy-plane where the velocity is zero. Geometrically describe the graph of this collection of points.

3. Suppose (x_0, y_0) is an eigenvector for A with eigenvalue $\lambda \neq 0$. Show that a particle initially at (x_0, y_0) will move in a straight line to the origin if $\lambda < 0$ and in a straight line away from the origin if $\lambda > 0$. (Straight line motion of our particle is the simplest motion. Such motions correspond to eigenvectors/eigenvalues of A.)

A Fish Problem

Suppose the population P for a rare species of fish in a lake is reasonably modeled by the differential equation $\dfrac{dP}{dt} = -P^2 + 100P$. It has been decided that fishing will be allowed in the lake, but how many fishing licenses to issue needs to be determined. Each fisherman with a license will be allowed to catch three fish per year.

• What is the largest number of licenses that can be issued so that the fish will have a chance of survival in the lake?

• Suppose the number of fishing licenses that you determined in the previous item is issued and that each licensed fisherman catches his three fish per year. What will eventually happen to the fish population?

10.3   The Tennis Serve

Objective

The goal of this project is to use differential equations to analyze the margin of error in a tennis serve.

Background

We assume the reader is familiar with the game of tennis. The server hits a ball over a net so as to land in the "fair" region of the court as pictured.

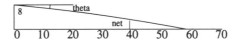

We assume the net is 3 feet high and is 39 feet from the server. In order for the serve to be valid, the ball must go over the net and land within 60 feet of the server.

Assumptions

We neglect air resistance on the ball and we assume the ball is hit "flat;" i.e., without a spin. (A spinning ball adds a "spin force" to the ball which makes analysis more complicated.) So the only force on the ball that we consider is the force due to gravity.

We assume the server hits the ball 8 feet above ground level and that the initial speed of the ball is 140 feet per second. (A speed of 140 feet per second is slightly over 95 miles per hour. Professional tennis players have serves this fast and even faster.) The initial velocity for the ball (a vector) has the form

$$\mathbf{v}_0 = 140\cos{(\theta)}\mathbf{i} + 140\sin{(\theta)}\mathbf{j},$$

where θ is the angle that the velocity vector makes with respect to the horizontal.

Problem

Our problem is to find those values of θ such that a tennis serve, with initial velocity $\mathbf{v}_0 = 140\cos{(\theta)}\mathbf{i} + 140\sin{(\theta)}\mathbf{j}$ and with initial position 8 feet above ground level, will be a "good serve;" i.e., go over the net and land within 60 feet of the server.

We use $g = 32$ ft/sec^2 as the gravitational constant, so $\mathbf{F} = -32m\mathbf{j}$, where m is the mass of the tennis ball. Let $\mathbf{r}(t) = x(t)\mathbf{i} + y(t)\mathbf{j}$ denote the location of the ball at time $t \geq 0$. We have $\mathbf{r}(0) = 8\mathbf{j}$ and $\mathbf{r}'(0) = 140\cos{(\theta)}\mathbf{i} + 140\sin{(\theta)}\mathbf{j}$. Recall that Newton's law says that the total force acting on the ball at time t is equal to the mass of the ball times its acceleration, $\mathbf{F}(t) = m\mathbf{r}''(t)$.

Instructions

Find values of α and β in radians such that if $\alpha < \theta \leq \beta$, then the serve is good. Do the same for θ measured in degrees. (Note that θ will be negative if the angle of the serve is below the horizontal.)

The tolerance you obtain on θ for the serve to be "in" is quite small. However the initial speed of the serve is at the professional level. As anyone who plays tennis knows, the slower the speed of the ball, the greater the tolerance for the angle θ and the easier it is for the serve to be "in."

If air resistance were taken into account in analyzing the tennis serve, would it be easier or harder to get the serve in? What about spin serves?

10.4   Modeling Lake Pollution

A manufacturing company has been discharging pollutant Y into Lake Pristine as part of an effluent that has concentration 10 grams of Y per cubic meter and that discharges at an average rate of 2×10^5 cubic meters per year. Rain water also flows into the lake at the rate of 3×10^6 cubic meters per year. The lake is controlled by a dam so that its volume stays at a constant 14.6×10^6 cubic meters. The Environmental Protection Agency has determined that the level of pollutant Y is too high and should be brought down to a concentration of 0.4 grams of Y per cubic meter within five years. Assume that the company decreases its concentration of pollutants linearly with time (i.e., the concentration is a linear function of t, where t is measured in years). Your job is to determine the *rate of decrease* of the concentration of pollutant in the effluent discharged by the company so that the concentration of pollutant Y in the lake will have dropped to the prescribed level in five years.

Suggestions

First compute the current concentration of Y in the lake. (This does not require solving a differential equation, just a little common sense.) Then use this concentration as an initial condition for a differential equation that models the concentration of the pollutant in the lake. Assume that the pollutants are well stirred after they enter the lake, and that lake water exits at a rate that equals the sum of the rates of entry of effluent and rain water into the lake. Your differential equation will involve the unknown rate of decrease of the concentration of the pollutant in the effluent. Use Maple to solve this differential equation (exactly). Solve for the rate of decrease of the concentration of the pollutant in the effluent by setting the concentration in the lake equal to 0.4 when $t = 5$.

Maple is not essential to do this project, but it does simplify the many computations. To keep track of the flow of ideas in your work use labels, such as V for the volume of the lake; and then substitute their values at the end for the final computations.

10.5   Shooting a Projectile

This project involves shooting a projectile vertically into the air, under various assumptions on the air resistance. The physical law that models projectile motion is Newton's Law, which reads

$$Force = mass \times acceleration.$$

When ignoring air resistance, the only force acting on the projectile is $-mg$ where m is the mass and g is the gravitational constant (the sign convention here assumes that the positive direction is up). If the velocity of the rocket is high, then air resistance cannot be ignored and is proportional to the square of its speed. In this scenario, the force term becomes $-mg - kv(t)^2$, where k is a proportionality constant and $v(t)$ is the velocity of the projectile at time t.

A more realistic version of this problem takes into consideration that air resistance decreases exponentially with altitude. In this scenario, the constant k in the above discussion would be replaced with $ke^{-\alpha x(t)}$ where α is a constant and $x(t)$ represents the altitude of the projectile at time t. The value of α can be computed, knowing that the air density at one mile (1.6 km) is roughly two-thirds the air density at sea level.

Your job is to solve the differential equation associated with each of the three sets of assumptions outlined above (i.e., without air resistance; with air resistance equal to kv^2; and with air resistance equal to $ke^{-\alpha x}v^2$). In all three scenarios, assume that the initial velocity is 0.5 kilometers per second. Remember to use metric units for the gravitational constant g. The differential equations governing the first two sets of assumptions can be solved exactly. The third can only be approximated numerically, say using Maple's dsolve with the numeric option. In all three cases, assume $m = 1$ kilogram. For the second and third scenarios, use $k = 0.01$. Use Maple's odeplot command (remember to execute with(plots); first) to view all three solutions. Compare the maximum heights of the projectile under each of the three scenarios. Which of the three solutions attains the smallest maximum height? Give a qualitative explanation as to why this particular solution should attain the smallest maximum height.

Part II of Project

The goal of this part of the project is to reexamine Section 8.5, taking into account that air resistance changes with altitude. Your job is to repeat the analysis, but with the constant k replaced by $ke^{-ay(t)}$, where the constant a is determined as above (i.e. the air density at 1 mile (1600 meters) is two-thirds the air density at sea level). Use the following values of the constants in Section 8.5.: km=0.0001, $g = 9.8$ meters per second per second, $v0 = 500$ meters per second, and $\alpha = \pi/4$.

The angle, 45 degrees ($= \pi/4$ radians), is close to, but not equal to, the angle which corresponds to the maximum horizontal distance traveled by the projectile. By adjusting alpha, find the angle (to the nearest degree) which corresponds to the maximum horizontal distance traveled by the projectile. Then, find the minimum initial velocity (v0) which is necessary to send the projectile a horizontal distance of 14,000 meters.

10.6   Rocket Propulsion

We consider a rocket of mass 500 kilograms including 300 kilograms of fuel. Assume that the fuel burns at a constant rate for a period of 60 seconds (at which point all the fuel has burned) and that the speed of the exhaust is 1500 meters per second (relative to the rocket). Assume that the rocket has zero initial velocity. Your job is to find the velocity of the rocket at $t = 60$ seconds under the following sets of assumptions.

1. *No air resistance.* The differential equation that models the rocket under the assumption of no air resistance is

$$-mg = m\frac{dv}{dt} - u\frac{dm}{dt},$$

   where m is the (variable) mass of the rocket, v is the velocity of the rocket, and u is the exhaust speed (relative to the rocket). Find the velocity of the rocket at 60 seconds.

2. *Air resistance.* Now assume that air resistance exerts a retarding force of $ke^{-\alpha x}v^2$, where α and k are constants, and x represents the altitude of the rocket. Take $k = 0.01$ and compute α knowing that the retarding force at 1.6 kilometers ($x = 1.6$) is two-thirds the retarding force at sea level. Modify the

force term on the left of the above differential equation to include this air resistance term (be careful to use the correct sign); and numerically approximate the solution using Maple's `dsolve` command with the `numeric` option. Find the velocity of the rocket 60 seconds after launch.

10.7   The Laplace Transform and Square Waves

Background

We wish to analyze a series RLC circuit having an applied voltage which is a square wave. Specifically, we would like to find the initial conditions which completely eliminate any transient in the circuit. The relevant initial value problem is

$$Lq'' + Rq' + \frac{1}{C}q = E(t), \quad q(0) = q_0, \quad q'(0) = q_1, \tag{10.4}$$

where $q = q(t)$ is the charge on the capacitor. We will assume that $L = 1$ henry, $R = 3$ ohm, $C = 1/2$ farad, and

$$E(t) = \left\{ \begin{array}{ll} 1 & \text{if } 0 \leq t < 1; \\ 0 & \text{if } 1 \leq t < 2, \end{array} \right.$$

with the square wave repeated periodically. The following commands show how to generate and plot a square wave with five periods.

```
>   n:=4:
>   f:=sum(Heaviside(t-2*k)-Heaviside(t-2*k-1),k=0..n);
```

$$f := \text{Heaviside}(t) - \text{Heaviside}(t - 1) + \text{Heaviside}(t - 2) - \text{Heaviside}(t - 3)$$
$$+ \text{Heaviside}(t - 4) - \text{Heaviside}(t - 5) + \text{Heaviside}(t - 6) - \text{Heaviside}(t - 7)$$
$$+ \text{Heaviside}(t - 8) - \text{Heaviside}(t - 9)$$

```
>   plot(f,t=0..12);
```

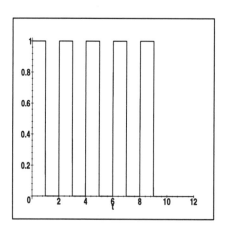

Instructions

For given initial conditions, solve (10.4) using Laplace transforms, and then plot the solution.

1. Solve (10.4) given that $q(0) = 0$, $q'(0) = 1$; then plot the solution in the (q, q')-plane. (*Hint:* You might want to use on-line help for the `plot` command to see how to make parametric plots.)

2. By trial and error, and by using the results of the preceding problem, find initial conditions for (10.4) for which the solution is periodic; i.e., there is no transient. Then plot the solution in the (q, q')-plane.

3. Find the exact initial values for which the solution to (10.4) is periodic.

10.8   Taylor Series Methods for Differential Equations

Background

Recall Euler's method for the initial value problem

$$\frac{dy}{dt} = f(t, y), \quad y(t_0) = y_0. \tag{10.5}$$

Pick a step size $h > 0$, then let $t_i = t_0 + ih$, $i = 0, 1, \ldots, n$, and let

$$y_{i+1} = y_i + hf(t_i, y_i), \ i = 0, 1, \ldots, n - 1, \tag{10.6}$$

where n is the number of steps of size h you must take to reach some prescribed final value of t. Euler's method, sometimes called the tangent line method, is based on a first degree Taylor series expansion of the solution $y(t)$ about $t = t_i$, assuming the solution goes through the point (t_i, y_i); i.e., $y(t_i) = y_i$:

$$y(t) = y(t_i) + y'(t_i)(t - t_i) + O((t - t_i))^2).$$

Letting $t = t_{i+1} = t_i + h$ and $y_{i+1} = y(t_{i+1})$ and keeping only the first two terms gives Euler's method (10.6). There is no reason we can't try a higher order Taylor expansion. If we use a second degree Taylor expansion, we get

$$y(t) = y(t_i) + y'(t_i)(t - t_i) + \frac{y''(t)(t - t_i)^2}{2} + O((t - t_i))^3),$$

which leads to the approximation

$$y_{i+1} = y_i + hy'(t_i) + \frac{h^2}{2}y''(t_i). \tag{10.7}$$

The differential equation (10.5) is used to find y' and y'' in (10.7). For example, if $y' = f(t, y) = t^2 + y^2$, then $y''(t) = \frac{d(t^2 + y^2)}{dt} = 2t + 2y(t)\frac{dy(t)}{dt} = 2t + 2y(t)(t^2 + y(t)^2)$ [Why?] You should verify this example with Maple's `diff` command. Be sure to replace y by $y(t)$ before differentiating $f(t, y(t))$ with respect to t.

Instructions

1. Verify that $y = 1 - \dfrac{2}{t}$ is the solution to the problem

$$\frac{dy}{dt} = f(t, y) = \frac{1-y}{t}, \quad y(1) = -1.$$

2. Use both Euler's method and the second order Taylor series method to approximate y(2) for the preceding problem, with step size $h = 1/4$, and compare your answers (called *Euler1* and *Taylor2*) with the exact solution, $y(2) = 0$.

3. Repeat Problem 2 with step sizes $h = 2^{-k}$, $k = 3, 4, 5, \ldots, 8$. Show that the error of Euler's method is approximately a constant times h, and that the error of the second order Taylor series method is approximately a constant times h^2.

4. Use the results of Problem 3 to estimate the step size required to obtain an accuracy of 5×10^{-9} for each method. How many steps would be required for each method to obtain this accuracy?

5. Repeat Problems 2 through 4 using the Taylor expansion of degree 4.

10.9   Controlling the Motion of a Cart

Suppose a supply cart is moving along a straight horizontal track next to an assembly line. The motion of the cart is determined by a motor which is capable of accelerating the cart at the rate of 1 m/sec^2 in either direction along the track. In addition, friction causes the cart to decelerate at a rate proportional to its speed, with proportionality constant k.

1. Write a differential equation which models the motion of the cart on the track. Justify the presence of each term.

2. Assume $k = 1/10$. Suppose that at time $t = 0$ the cart is positioned 20 meters to the right of Station A and is moving to the left at the rate of v_0 m/sec with $v_0 \geq 0$. Station B is positioned 40 meters to the right of Station A. Give a precise formulation of how you can use the acceleration capabilities of the cart to bring it to Station B as quickly as possible, and have it arrive with zero velocity. Give a plot of v_0 versus arrival time.

3. Now consider the case for general nonnegative k. Fix $v_0 = 20$ m/sec and give a plot of k versus arrival time. Does friction help or hinder the arrival? Given that the cart can never be controlled as precisely as in the mathematical model, what vital role will friction play in the process?

We discuss the solutions to 1 and 2 below. After studying these, you are to solve 3.

Solutions

1. Pick a point along the track and denote it as zero. Then treat the track as a number line, with points to the right of zero denoted in terms of positive meters and points to the left of zero denoted in terms of

negative meters. Let the position of the cart at time t be given by $s(t)$. Then the motion of the cart is governed by the differential equation

$$s''(t) = -ks'(t) + u(t)$$

where $u(t)$ can take on the values -1, 0, or 1. The term $s''(t)$ is the acceleration of the cart at time t. The differential equation indicates that the acceleration of the cart decomposes into the terms $-ks'(t)$ and $u(t)$. The term $-ks'(t)$ is due to friction, and the term $u(t)$ is due to the activity or non-activity of the motor.

2. Take Station A as the zero point along the track. Then Station B is located at the point 40 along the track. We are given that $k = 1/10$, $s(0) = 20$, and $s'(0) = -v_0$. We want to describe how to assign values to $u(t)$ so that there exists $T > 0$ such that $s(T) = 40$, $s'(T) = 0$, and T is as small as possible.

Since the cart is initially moving to the left and we want the cart ultimately to be at B (to the right), we should begin by switching the motor to apply acceleration to the right. This action will ultimately slow the cart and reverse its motion. We will then need to determine at which point the motor should be switched to apply acceleration to the left so that we decelerate and arrive at Station B with zero velocity. Of course, as soon as this occurs, we will need to switch the motor off. As a result, our function $u(t)$ will have the form

$$u(t) = \begin{cases} 1 & 0 \le t < T_1, \\ -1 & T_1 \le t < T, \\ 0 & t \ge T, \end{cases}$$

where T_1 is the switching time from right to left acceleration. Consequently, we begin by solving the initial value problem

$$s'' + \tfrac{1}{10}s' = 1,$$
$$s(0) = 20,$$
$$s'(0) = -v_0,$$

to obtain

$$s(t) = 10t - 80 - 10v_0 + (100 + 10v_0)e^{-\frac{1}{10}t}.$$

Then

$$s(T_1) = 10T_1 - 80 - 10v_0 + (100 + 10v_0)e^{-\frac{1}{10}T_1}$$

and

$$s'(T_1) = 10\left(1 - e^{-\frac{1}{10}T_1}\right).$$

So we need to solve the initial value problem

$$s(T_1) = 10T_1 - 80 - 10v_0 + (100 + 10v_0)e^{-\frac{1}{10}T_1},$$
$$s'(T_1) = 10\left(1 - e^{-\frac{1}{10}T_1}\right),$$
$$s'' + \frac{1}{10}s' = -1.$$

The exact solution is given by

$$s(t) = -10t + 20T_1 + 10e^{-\frac{1}{10}T_1}v_0 + 120 - 10v_0 + 100\,\frac{-2 + e^{-\frac{1}{10}T_1}}{e^{-\frac{1}{10}T_1}}e^{-\frac{1}{10}t}.$$

We evaluate this expression at $t = T$ and set it equal to 40, (corresponding to arriving at Station B). We evaluate the derivative at $t = T$ and set it equal to 0 (corresponding to arriving with zero velocity). This yields the system of equations

$$-10T + 20T_1 + 10e^{-\frac{1}{10}T_1}v_0 + 120 - 10v_0 + 100\frac{-2+e^{-\frac{1}{10}T_1}}{e^{-\frac{1}{10}T_1}}e^{-\frac{1}{10}T} = 40,$$

$$-10 - 10\frac{-2+e^{-\frac{1}{10}T_1}}{e^{-\frac{1}{10}T_1}}e^{-\frac{1}{10}T} = 0.$$

Solving (2) for T_1 yields

$$T_1 = -10\ln\left(-20\frac{e^{-\frac{1}{10}T}}{-10 - 10e^{-\frac{1}{10}T}}\right).$$

Substituting this into (2) yields a *very nasty relation* between T and v_0. The plot of arrival time T versus initial speed v_0 (toward the left) follows.

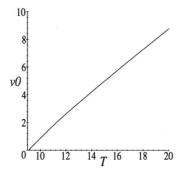

Chapter 11

Appendix.

11.1   Nuts and Bolts

This is a compendium of all the things that your mother wanted to protect you from. It delves into Maple's programming language more deeply than many first time readers need to do. If you are content to work only with the carefully contrived examples in most books, the ones whose answers are all integers, it is probably overkill. It was the original intention of the authors to avoid most of these messy details, but students and faculty kept coming up with questions that cannot be answered without confronting that particular devil. Here, then is the unexpurgated version.

11.1.1   Culling Eigenvectors

The alert reader will have noticed that, as in all good mathematical exposition, a certain benevolent deceit has been perpetrated. In particular, the examples in earlier sections acknowledged that eigenvectors are seldom nice: the best that one can do is to get floating point decimal approximations. However, the examples that were studied in detail all had integer entries in their eigenvectors, so that it was easy to retype them into the single column matrices that were used to generate solutions. What does one do if the eigenvectors and eigenvalues are all messy floating point decimals? One of the things that we will need to do often in studying differential equations is to take a matrix and pick out one or more eigenvectors. Maple has a command eigenvects in the linalg package that will do this, but one needs to know just how to use it. In particular, if the matrix has integer entries, then the command will try to return the characteristic polynomial in the form of RootOf, which is useless to us. The easiest way to avoid this problem is to convert the entries to floating point decimals by means of the map command, which applies a function to each element of an array.

```
>  with(linalg):

Warning, new definition for norm


Warning, new definition for trace


>  randmatrix(3,3);
```

$$\begin{bmatrix} -35 & 97 & 50 \\ 79 & 56 & 49 \\ 63 & 57 & -59 \end{bmatrix}$$

```
>  A:=map(evalf,");
```

$$A := \begin{bmatrix} -35. & 97. & 50. \\ 79. & 56. & 49. \\ 63. & 57. & -59. \end{bmatrix}$$

We can then compute the eigenvectors, using `eigenvects`. Note that the eigenvectors will seldom be integers; and complex numbers occur often, even if the original matrix has all real entries.

```
>  s:=eigenvects(A);
```

$$s := [-69.22606858, 1, \{[-.2905580334, .4378534965, -.6505425980]\}],$$
$$[-106.3997066, 1, \{[-.7758514404, .1040190166, .9061144082]\}],$$
$$[137.6257748, 1, \{[-.565504967, -.7943096200, -.4114539984]\}]$$

Our first problem is to pick the eigenvectors from this forest of data. Since the eigenvectors are not simple integers, we would prefer not to have to copy them digit by digit.

The output is in the form of an expression sequence, with individual entries separated by commas. We can pull out the jth expression by using j as a subscript.

```
>  s[2];
```

$$[-106.3997066, 1, \{[-.7758514404, .1040190166, .9061144082]\}]$$

The first number is the eigenvalue, the second is the multiplicity (how often it is a root of the characteristic equation), and the last entry is a set containing our eigenvector.

We can pull out the set by asking for the third subscript, and we can get rid of the curly braces by using the op command.

```
>  s[2][3];
```

$$\{[-.7758514404, .1040190166, .9061144082]\}$$

```
>  v:=op(");
```

$$v := [-.7758514404, .1040190166, .9061144082]$$

Thus, we have isolated the second eigenvector. However, we would prefer to see it as a vertical object, rather than a horizontal one. Since Maple's `transpose` does not change horizontal objects (vectors, row vectors of a matrix, column vectors of a matrix) into vertical objects, we have to do it on our own. First we change the vector into a list, I then we make the list into a matrix with one column.

```
>  convert(",list);
```

$$[-.7758514404, .1040190166, .9061144082]$$

```
>  matrix(vectdim("),1,");
```

$$
\begin{bmatrix}
-.7758514404 \\
.1040190166 \\
.9061144082
\end{bmatrix}
$$

Since we will often have to do this operation, we can write a Maple procedure to automate the process:

```
>  cm:=proc(x)
   if not (type(x,list) or type(x,vector)) then
   ERROR('expecting a list or a vector') fi;
   convert(eval(x),list):
   matrix(vectdim(x),1,");
   end:

>  cm(v);
```

$$
\begin{bmatrix}
-.7758514404 \\
.1040190166 \\
.9061144082
\end{bmatrix}
$$

In fact, we can also automate the selection of an eigenvector.

```
>  svec:=proc(x,y)
   if args[nargs] >=nargs then
   ERROR('eigenvalue index too large') fi:
   if whattype(s)<>exprseq then
   ERROR('expecting an expression sequence') fi:
   op(args[args[nargs]][3]);
   end:
   cm(svec(s,2));
```

$$
\begin{bmatrix}
-.7758514404 \\
.1040190166 \\
.9061144082
\end{bmatrix}
$$

In fact, we can make a procedure to place the eigenvectors as columns in a matrix.

```
>  evects:=proc(x)
   local j;
   augment(seq(cm(svec(args,j)),j=1..nargs)):
   end:

>  evects(s);
```

$$
\begin{bmatrix}
-.2905580334 & -.7758514404 & -.565504967 \\
.4378534965 & .1040190166 & -.7943096200 \\
-.6505425980 & .9061144082 & -.4114539984
\end{bmatrix}
$$

The same approach can be used to select the eigenvalues. (Note that one cannot use the `eigenvals` command, since the order in which the eigenvalues are selected will likely not be the same.)

```
> sval:=proc(x,y)
  if args[nargs] >=nargs then
  ERROR('eigenvalue index too large') fi:
  if whattype(s)<>exprseq then
  ERROR('expecting an expression sequence') fi:
  args[args[nargs]][1];
  end:
> evals:=proc(x)
  local i;
  [seq(sval(args,i),i=1..nargs)];
  end:
> evals(s);
```

$$[-69.22606858, -106.3997066, 137.6257748]$$

The reader should note that Maple also has a command `Eigenvals( ,vecs)` that will do something similar in one step. However, one has to be very careful:

1. One has to unassign the name of the matrix that the vectors will appear in before the operation is commenced,

2. One has to have a complex valued matrix to begin with (if the matrix is real and the eigenvalues are complex, Maple has a strange output form),

3. One has to use the evaluated form of the matrix, rather than just the name, and

4. The command is inert, i.e., it does nothing until the result is passed to `evalf`.

```
> vec:='vec';
```

$$vec := vec$$

```
> B:=A*(1+0.0000000001*I);
```

$$B := (1. + .1\,10^{-9}\,I)\,A$$

```
> evalm(");
```

$$\begin{bmatrix} -35. - .35\,10^{-8}\,I & 97. + .97\,10^{-8}\,I & 50. + .50\,10^{-8}\,I \\ 79. + .79\,10^{-8}\,I & 56. + .56\,10^{-8}\,I & 49. + .49\,10^{-8}\,I \\ 63. + .63\,10^{-8}\,I & 57. + .57\,10^{-8}\,I & -59. - .59\,10^{-8}\,I \end{bmatrix}$$

```
> Eigenvals(",vec):
> evalf(");
```

$$[137.6257751 + .1376257753\,10^{-7}\,I, -106.3997064 - .1063997064\,10^{-7}\,I,$$
$$-69.22606861 - .6922606856\,10^{-8}\,I]$$

```
>  V:=evalm(vec);
```

$V :=$
$$[.5343476652 + .2671738326\ 10^{-9}\ I,\ -.6479935375 - .3878009110\ 10^{-9}\ I,$$
$$.3681741463 + .2209046351\ 10^{-9}\ I]$$
$$[.7505460001 + .3752730002\ 10^{-9}\ I,\ .08687700729 + .5199277559\ 10^{-10}\ I,$$
$$-.5548163161 - .3328900118\ 10^{-9}\ I]$$
$$[.3887843539 + .1943921771\ 10^{-9}\ I,\ .7567895705 + .4529114378\ 10^{-9}\ I,$$
$$.8243205832 + .4945926795\ 10^{-9}\ I]$$

The reader should note that the columns given by the last procedure are different from the columns from the first procedure. This is to be expected, since any scalar multiple of an eigenvector is again an eigenvector. From the above matrix V, we can pick off individual eigenvectors using the col command and the procedure cm. For example, to pick off the first column, we input

```
>  col(V,1);
```

$$[.5343476652 + .2671738326\ 10^{-9}\ I,\ .7505460001 + .3752730002\ 10^{-9}\ I,$$
$$.3887843539 + .1943921771\ 10^{-9}\ I]$$

```
>  cm(");
```

$$\begin{bmatrix} .5343476652 + .2671738326\ 10^{-9}\ I \\ .7505460001 + .3752730002\ 10^{-9}\ I \\ .3887843539 + .1943921771\ 10^{-9}\ I \end{bmatrix}$$

Alternatively, we can get the first column with submatrix.

```
>  submatrix(V,1..3,1..1);
```

$$\begin{bmatrix} .5343476652 + .2671738326\ 10^{-9}\ I \\ .7505460001 + .3752730002\ 10^{-9}\ I \\ .3887843539 + .1943921771\ 10^{-9}\ I \end{bmatrix}$$

11.1.2   Systems and Matrices

How does one go back and forth from systems to matrices? Suppose that we have a coefficient matrix A that is related to a system of n simultaneous differential equations. How can one solve the associated system by using dsolve, for example to compare with the results of a matrix solution?

```
>  A:=matrix(3,3,[-6,2,2,-13,-4,1,15,-5,-10]);
```

$$A := \begin{bmatrix} -6 & 2 & 2 \\ -13 & -4 & 1 \\ 15 & -5 & -10 \end{bmatrix}$$

```
>  Y:=[seq(y.i(x),i=1..3)];
```

$$Y := [\text{y1}(x),\ \text{y2}(x),\ \text{y3}(x)]$$

```
>  i:='i':
>  matrix(3,1,Y);
```

$$\begin{bmatrix} \text{y1}(x) \\ \text{y2}(x) \\ \text{y3}(x) \end{bmatrix}$$

```
>  map(diff,",x)-A &* ";
```

$$\begin{bmatrix} \frac{\partial}{\partial x}\,\text{y1}(x) \\ \frac{\partial}{\partial x}\,\text{y2}(x) \\ \frac{\partial}{\partial x}\,\text{y3}(x) \end{bmatrix} - (A\,\&*\ \begin{bmatrix} \text{y1}(x) \\ \text{y2}(x) \\ \text{y3}(x) \end{bmatrix})$$

```
>  evalm(");
```

$$\begin{bmatrix} (\frac{\partial}{\partial x}\,\text{y1}(x)) + 6\,\text{y1}(x) - 2\,\text{y2}(x) - 2\,\text{y3}(x) \\ (\frac{\partial}{\partial x}\,\text{y2}(x)) + 13\,\text{y1}(x) + 4\,\text{y2}(x) - \text{y3}(x) \\ (\frac{\partial}{\partial x}\,\text{y3}(x)) - 15\,\text{y1}(x) + 5\,\text{y2}(x) + 10\,\text{y3}(x) \end{bmatrix}$$

```
>  eqseq:=seq("[i,1],i=1..3);
```

$$eqseq := (\frac{\partial}{\partial x}\,\text{y1}(x)) + 6\,\text{y1}(x) - 2\,\text{y2}(x) - 2\,\text{y3}(x),\ (\frac{\partial}{\partial x}\,\text{y2}(x)) + 13\,\text{y1}(x) + 4\,\text{y2}(x) - \text{y3}(x),$$
$$(\frac{\partial}{\partial x}\,\text{y3}(x)) - 15\,\text{y1}(x) + 5\,\text{y2}(x) + 10\,\text{y3}(x)$$

```
>  i:='i':
>  sol:=dsolve({eqseq},Y);
```

$$sol := \{\text{y1}(x) = \frac{2}{5}\,\_C1\,e^{(-5\,x)} - \frac{2}{5}\,\_C1\,e^{(-10\,x)} + \frac{1}{5}\,\_C2\,e^{(-10\,x)} + \frac{4}{5}\,\_C2\,e^{(-5\,x)}$$
$$+ \frac{2}{5}\,\_C3\,e^{(-5\,x)} - \frac{2}{5}\,\_C3\,e^{(-10\,x)}, \text{y2}(x) = -\frac{6}{5}\,\_C1\,e^{(-10\,x)} + \frac{11}{5}\,\_C1\,e^{(-5\,x)} - 5\,\_C1\,x\,e^{(-5\,x)}$$
$$- 10\,\_C2\,x\,e^{(-5\,x)} - \frac{3}{5}\,\_C2\,e^{(-5\,x)} + \frac{3}{5}\,\_C2\,e^{(-10\,x)} - 5\,\_C3\,x\,e^{(-5\,x)} + \frac{6}{5}\,\_C3\,e^{(-5\,x)}$$
$$- \frac{6}{5}\,\_C3\,e^{(-10\,x)}, \text{y3}(x) = -2\,\_C1\,e^{(-5\,x)} + 2\,\_C1\,e^{(-10\,x)} + 5\,\_C1\,x\,e^{(-5\,x)}$$
$$+ 10\,\_C2\,x\,e^{(-5\,x)} + \_C2\,e^{(-5\,x)} - \_C2\,e^{(-10\,x)} + 2\,\_C3\,e^{(-10\,x)} + 5\,\_C3\,x\,e^{(-5\,x)}$$
$$- \_C3\,e^{(-5\,x)}\}$$

Suppose that we have a list L containing n simultaneous differential equations in the variables $y_1(x)$, $y_2(x)$, ..., $y_n(x)$ with the derivatives on the left hand side and the variables on the right hand side. How does one extract the matrix A?

```
> diffeq1:=diff(y1(x),x)=-6*y1(x)+2*y2(x)+2*y3(x);
```

$$diffeq1 := \frac{\partial}{\partial x} y1(x) = -6\,y1(x) + 2\,y2(x) + 2\,y3(x)$$

```
> diffeq2:=diff(y2(x),x)=-13*y1(x)-4*y2(x)+y3(x);
```

$$diffeq2 := \frac{\partial}{\partial x} y2(x) = -13\,y1(x) - 4\,y2(x) + y3(x)$$

```
> diffeq3:=diff(y3(x),x)=15*y1(x)-5*y2(x)-19*y3(x);
```

$$diffeq3 := \frac{\partial}{\partial x} y3(x) = 15\,y1(x) - 5\,y2(x) - 19\,y3(x)$$

```
> L:=[diffeq1,diffeq2,diffeq3];
```

$$L := [\frac{\partial}{\partial x} y1(x) = -6\,y1(x) + 2\,y2(x) + 2\,y3(x),\ \frac{\partial}{\partial x} y2(x) = -13\,y1(x) - 4\,y2(x) + y3(x),$$
$$\frac{\partial}{\partial x} y3(x) = 15\,y1(x) - 5\,y2(x) - 19\,y3(x)]$$

First, we form the list of variables, then we use `genmatrix`.

```
> Y:=[seq(y.i(x),i=1..3)];
```

$$Y := [y1(x),\ y2(x),\ y3(x)]$$

```
> i:='i':
```

```
> A:=genmatrix(map(rhs,L),Y);
```

$$A := \begin{bmatrix} -6 & 2 & 2 \\ -13 & -4 & 1 \\ 15 & -5 & -19 \end{bmatrix}$$

Higher Order Equations to Systems

Suppose that we have a differential equation of order greater than one with characteristic polynomial p. How does one write the matrix for the system of simultaneous first order equations that results when one sets $y_1(t) = y(t)$, $y_2(t) = y'(t)$, $y_3(t) = y''(t)$, etc? For example, consider the equation

$$y'''' - 2y''' - 3y'' + 5y' - 7y = 0.$$

First we form the characteristic equation, then we transpose the companion matrix.

```
> p:=x^4-2*x^3-3*x^2+5*x+7;
```

$$p := x^4 - 2\,x^3 - 3\,x^2 + 5\,x + 7$$

```
> companion(p,x);
```

$$
\begin{bmatrix}
0 & 0 & 0 & -7 \\
1 & 0 & 0 & -5 \\
0 & 1 & 0 & 3 \\
0 & 0 & 1 & 2
\end{bmatrix}
$$

```
> transpose(");
```

$$
\begin{bmatrix}
0 & 1 & 0 & 0 \\
0 & 0 & 1 & 0 \\
0 & 0 & 0 & 1 \\
-7 & -5 & 3 & 2
\end{bmatrix}
$$

The reader is also invited to examine the Release 4 command `convertsys` in the `DEtools` package.

11.1.3   Data Types and Sizes

Enterprising students (that includes faculty members just learning Maple) often want to write procedures to automate these routines. The first thing that must be mastered for such a procedure to work correctly is the ability to distinguish these different linear algebraic objects and to determine their sizes. Maple has numerous data types. An object can be a list, a listlist, an array, a matrix, or a vector. They can sometimes look alike, but react differently in commands. Note that the object gotten by extracting a row or a column from a matrix is written *horizontally*, as are vectors.

```
> with(linalg):
> L:=[x,x^2,y];
```

$$
L := [x, x^2, y]
$$

```
> LL:=[L,L]; #the lists
```

$$
LL := [[x, x^2, y], [x, x^2, y]]
$$

```
> A:=array(1..2,1..2,[[x,x^2],[y,y^2]]);
```

$$
A := \begin{bmatrix} x & x^2 \\ y & y^2 \end{bmatrix}
$$

```
> M:=matrix(2,2,[x,x^2,y,y^2]);
```

$$
M := \begin{bmatrix} x & x^2 \\ y & y^2 \end{bmatrix}
$$

```
>  COLM:=matrix(3,1,[x,x^2,y]); #the arrays
```

$$COLM := \begin{bmatrix} x \\ x^2 \\ y \end{bmatrix}$$

```
>  V:=vector(2,[x,x^2]);
```

$$V := [x, \, x^2]$$

```
>  MC:=row(M,1);
```

$$MC := [x, \, x^2]$$

```
>  MR:=col(M,1); #the vectors
```

$$MR := [x, \, y]$$

For each of these data types, one can ask whattype, which returns a word for the kind of object that the argument is. Both a list and a listlist yield list. An array, a matrix, and all the types of vectors return array, but *only* after they have been evaluated using evalm. Until they are evaluated, they return string.

```
>  whattype(LL);
```

list

```
>  whattype(M);
```

string

```
>  whattype(evalm(M));
```

array

In addition to asking whattype, which returns a type, one can also ask type(,datatype), where the type can be list, array, matrix, or vector (among others). The second kind of query returns true or false. As one would expect, it is true that an array is a matrix and a matrix as an array. (Technically, an array is a matrix only if the indices of the array entries start with one: you can index an array starting at -2 if desired.) A listlist is neither a matrix nor an array. A matrix with only one row or one column returns false, but the object that one gets from a matrix by extracting a row or column returns true. All lists return false. In summary, whattype is useful to distinguish lists from arrays, but the type(,datatype) is needed to distinguish matrices from vectors.

```
>  type(A,matrix);
```

true

```
>  type(M,array);
```

true

```
>  type(LL,matrix);
```

false

```
>   type(LL,array);
```
$$false$$

```
>   type(COLM,vector);
```
$$false$$

```
>   type(MR,vector);
```
$$true$$

```
>   type(MC,vector);
```
$$true$$

The subs command works as expected with lists. However, the subs command will work only after an array object has been evaluated.

```
>   subs(x=2,LL);
```
$$[[2,\ 4,\ y],\ [2,\ 4,\ y]]$$

```
>   evalm(subs(x=2,A));
```
$$\begin{bmatrix} x & x^2 \\ y & y^2 \end{bmatrix}$$

```
>   subs(x=2,evalm(A));
```
$$\begin{bmatrix} 2 & 4 \\ y & y^2 \end{bmatrix}$$

```
>   evalm(subs(x=2,V));
```
$$[x,\ x^2]$$

```
>   subs(x=2,evalm(V));
```
$$[2,\ 4]$$

```
>   evalm(subs(x=2,MR));
```
$$[x,\ y]$$

```
>   subs(x=2,evalm(MR));
```
$$[2,\ y]$$

The command rowdim can be used the measure the size of arrays, matrices, and listlists.

```
>   rowdim(A);
```

```
>  rowdim(M);
```
$$2$$

```
>  rowdim(COLM);
```
$$3$$

```
>  rowdim(LL);
```
$$2$$

Similarly, `coldim` works for matrices with one column or listlists, but fails with lists.

```
>  coldim(COLM);
```
$$1$$

```
>  coldim(LL);
```
$$3$$

```
>  coldim(L);
```
```
Error, (in coldim) expecting a matrix
```

There is a command `vectdim` which can be used to measure the size of lists and vectors, but which does not work with a matrix of only one row or column (for which `rowdim` or `coldim` can be used, respectively.)

```
>  vectdim(L);
```
$$3$$

```
>  vectdim(LL);
```
$$2$$

```
>  vectdim(V);
```
$$2$$

```
>  vectdim(MC);
```
$$2$$

```
>  vectdim(MR);
```
$$2$$

```
>  vectdim(COLM);
```
```
Error, (in vectdim) expecting a vector
```

11.2   A Tridiagonal Solver

Although Maple's linear equation solvers are designed to perform well with large sparse matrices such as the tridiagonal matrices, we include the following routine specifically designed for the banded matrix situation encountered in the finite difference approach.

The equations (8.4) form a linear system of algebraic equations of the form

$$\mathbf{A}\mathbf{u} = \mathbf{b}, \tag{11.1}$$

where $\mathbf{A}$ is a tridiagonal $n \times n$ matrix. Such a system is readily solved, for example by Gaussian elimination, as illustrated in the following Maple procedure, called `trid`. The one-dimensional arrays **sb, d, sp** stand for the subdiagonal, diagonal, and superdiagonal entries in $\mathbf{A}$, while **b** contains the right hand side of (8.5) on input and the solution vector **u** on output. We will illustrate the use of `trid` to solve the problem (8.2) by the finite difference method, and then compare our numerical solution to the exact solution. The maximum relative error is seen to be around 0.01%.

```
>   restart:

>   EI:=10^9:L:=300:w:=8:P:=10^5:

>   diffeq:=(D@@2)(u)(x)+P*u(x)/EI=w*x*(L-x)/(2*EI):

>   sol:=dsolve({diffeq,u(0)=0,u(L)=0},u(x)):

>   uexact:=rhs(sol):

>   trid:=proc(n,sb,d,sp,b)
    local i;
    for i from 2 to n do
    sb[i]:=sb[i]/d[i-1]:
    d[i]:=d[i]-sp[i-1]*sb[i]:
    b[i]:=b[i]-b[i-1]*sb[i]:
    od:
    b[n]:=b[n]/d[n]:
    for i from n-1 by -1 to 1 do
    b[i]:=(b[i]-b[i+1]*sp[i])/d[i]:
    od:
    RETURN (sb,d,sp,b):
    end:

>   n:=299:sb:=array(1..n):d:=array(1..n):   # Note h=1.

>   sp:=array(1..n):b:=array(1..n):

>   PEI:=evalf(P/EI):wEI:=evalf(w/(2*EI)):

>   for i to n do
    sb[i]:=1:
    d[i]:=PEI-2:
    sp[i]:=1:
    b[i]:=wEI*i*(L-i):
    od:

>   soln:=trid(n,sb,d,sp,b):

>   usol:=evalm(soln[4]):
```

```
>   maxrelerr:=0:

>   for i to n do
    ux:=evalf(subs(x=i,uexact));
    relerr:=abs((usol[i]-ux)/ux):
    if relerr>maxrelerr then maxrelerr:=relerr:  fi:
    od:

>   maxrelerr;
```

$$.0001022503042$$

Index